In Memory

John Crim of New Market, VA

June 10, 1993

I sure do miss you, John. Your talent for cutting stones, your smile, that impish twinkle in your eye, and that willingness to do any job, anywhere, anytime are truly missed, very deeply!

It was a privilege and an honor to have known you. I only regret that it was for such a short period of time. The void that is left by your absence is felt every day.

Until we meet again . . .

E.S.

Moods In Brass & Glass is a new and different approach to wirewrapping. Emphasis is on the less expensive materials as the name implies. This volume also addresses two important needs: It is a creative supplement to the present day trend of "I want to do it myself", and a relatively inexpensive introduction to affordable materials, enabling one to become involved on a much deeper level without a lot of expense. In this book there is a strong emphasis on single and multi-bead projects as well as wire-only pendants, rings, and earrings. Each project is introduced, systematically diagrammed step by step, then summarized by suggesting variations of what has been covered. Further emphasis is placed upon creativity. This book, like *MOODS IN WIRE* Volume 1, will put you on a strong creative threshold so you can begin your own innovative journey in whatever direction you choose. The strong undercurrent of learning to "play with the wire" forces you to really discover yourself and your own capabilities.

Moods In Brass & Glass™

A Supplementary Guide
to the Fine Art
of Wirewrapping

VOLUME 2

Ellsworth "Ed" Sinclair

Published by E.E. Sinclair
P.O. Box 2011
Manassas, VA 22110

ISBN 0-9640483-1-0

Library of Congress Catalog
Number 95-96251

First printing 1996

Project Coordination and Printing:
Johnson Printing, Loveland, CO

Cover and Insert Photography:
Joyce Jay, Denver, CO

Cover/Insert Art Direction and Design:
LaVonne Ewing, LaSalle, CO

Electronic Publishing:
Teel & Co., Loveland, CO

Computer Illustrations:
William W. Ivers, Denver, CO

Portrait of Mr. Sinclair:
James L. Digby, Loveland, CO

ELLSWORTH "ED" SINCLAIR

Ed Sinclair

A man of diverse interests, Ed Sinclair studied drama at the Shakespearean Institute at Stratford-on-Avon in 1959, and followed that with a two year stint with the Peace Corps in Liberia, West Africa, where he "taught a little bit of everything, including history, literature and carpentry." He recently retired from a twenty-five year teaching career in Fairfax County, Virginia. Ed Sinclair is a graduate of the University of East Carolina and the University of Virginia.

As a native Virginian, he has spent many hours pursuing his mineral collecting hobby throughout the state. Early experiences involved being Field Trip Chairman (and later President) of the Mineralogical Society of the District of Columbia. He is currently a member of the Woodbridge Mineral Club and the Northern Virginia Mineral Club. He still collects when time permits. His interest in minerals led him to frequent the gem and mineral shows where he was introduced to wirewrapping. After a short 10 hour course and a year of practice, he embarked upon the craft show circuit in the mid-1970s, doing craft shows from Gettysburg, Pennsylvania to Asheville, North Carolina. Also, at that time, he served two terms as President of the Northern Virginia Handcrafters Guild, of which he is still an active member.

Ed lives in Manassas with his wife and son and spends many weekends on the craft show circuit. Future plans include retiring to the Shenandoah Valley near Harrisonburg.

Preface

Wirewrapping is a widely practiced art. Many systems are employed by many wirewrappers. Expressed in this book are many systems of my own, plus variations I have seen or learned.

It is not my intention in this book to copy, duplicate, plagiarize, or infringe upon any existing copyrighted material. Any resemblance to such an attempt is purely coincidental and unintentional.

Dedication

Volume 2 is dedicated to all those wonderful people out there who have made Volume 1, *MOODS IN WIRE,* so successful. Your interest, comments, and suggestions have been very rewarding and have helped make Volume 2 possible.

It is fitting at this time to recognize a new personality emerging on the wirewrap scene. I especially want to recognize a colleague of mine, Mr. Grant Hodges, whose single bead earring designs appear in this volume. He is also responsible for the left-hand-right-hand disclaimer in Volume 1, and currently he is my "left hand advisor". We will all be hearing more from him in the future.

TABLE OF CONTENTS

INTRODUCTION

Welcome to a new dimension in the wonderful world of wirewrapping where there is no end to variety. I am still amazed at the things people bring me to wrap. Included in this list is everything from a hand carved peach pit, to sea shells, to branches of coral, tumbled glass, and a $20 gold piece. From this variety comes much of the inspiration for the pages that follow. Since these objects are a far cry from the world of semi-precious stones, why not wrap them with a more appropriate, inexpensive material such as BRASS or COPPER? Hence, the name *BRASS & GLASS.*

Also included in this book will be glass beads and cabochons, very inexpensive acrylic cabochons that can be purchased at your local craft store.

By using this book, virtually everyone can be assured of having a good collection of very attractive costume jewelry. Please keep in mind that the instructions in this volume can also be used for gold and silver wire and semi-precious stones, and in like manner, Volume 1 can also be used on systems explained in this book. If Volume 2 is your first venture into wirewrapping, it is appropriate at this time to quote some advice from Volume 1:

- When wrapping, do not rush. If you are in a hurry to finish, then wirewrapping is not for you.

- **PATIENCE, PERSEVERANCE AND PRACTICE IS THE NAME OF THE GAME.** It is absolutely imperative to do your homework. At least one wrap a day will help you develop the necessary feel for the wire and will help keep the fingers nimble. Do not be too critical. Do not contemplate a piece too long. "Learn to do by doing and doing... and doing... and doing. Above all, don't ever give up... **PERSEVERE**!"

- Wirewrapping is like opening the door to one's imagination. The possibilities are limitless. Be willing to accept any challenge, to walk that path not often traveled before. Dare to be different, and you too will be a "happy wrapper."

- Some procedures in this book are covered in Volume 1, and it may be necessary to refer to that text before beginning projects described in Volume 2.

- Remember: **"PATIENCE, PERSEVERANCE AND PRACTICE IS THE NAME OF THE GAME."**

Instructions are written for right-handed people; if you are left-handed, please reverse the instructions accordingly.

CHAPTER 1 BASIC TOOLS FOR WIREWRAPPING

#1 EQUIPMENT

Pliers: All pliers should be the 4" size.

Chain Nose Pliers: These pliers are similar to the tapered round nose pliers except these pliers have a flat surface on the inside of the jaws. These pliers are used to tuck in ends of wires. They provide a real good grip to forcefully handle the wire without scratching it. They are also used to force the end of the wire through extremely tight places.

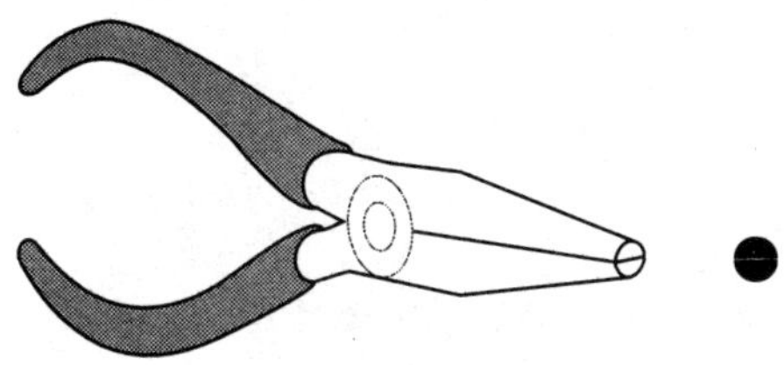

Chain Nose Pliers

Flat Nose Pliers: The pliers are used to bend over ends of wrap wires, crimp ends together, hold wires while bending, straighten bent wires, flatten wire, and hold groups of wires flat.

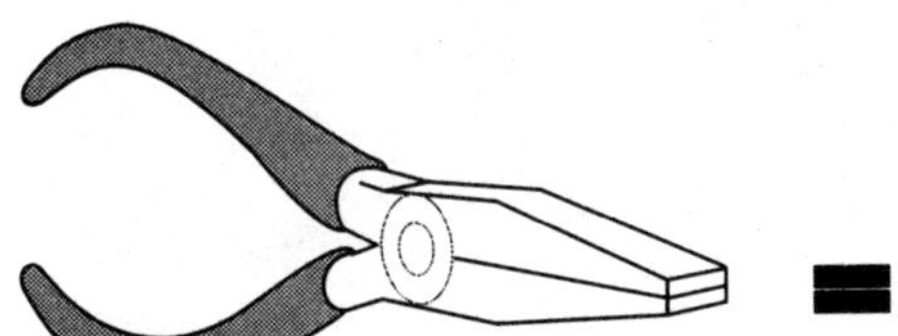

Flat Nose Pliers

Round Nose Pliers: These pliers are similar to the chain nose pliers except these are tapered a little more and have no flat surface to grip the wire. These pliers are used to make various sizes of loops. They can also be used to tuck in ends and are very useful when making earrings.

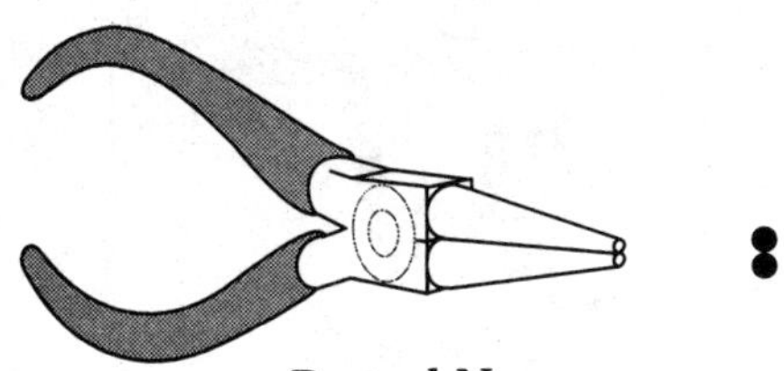

Round Nose Pliers

Side Cutters: Used for cutting the wire.

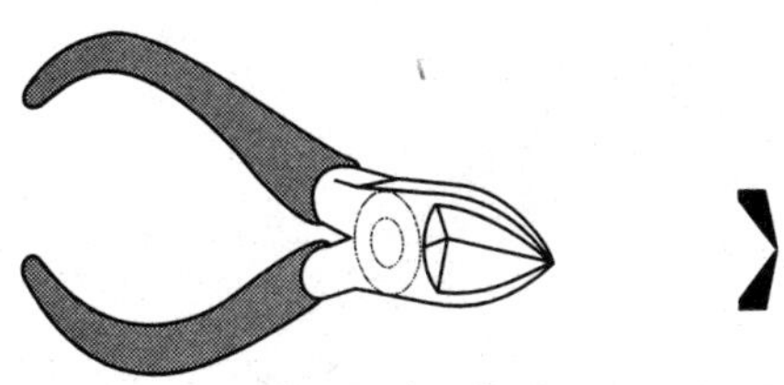

Side Cutters (Semi-Flush)

CHAPTER 1 BASIC TOOLS FOR WIREWRAPPING

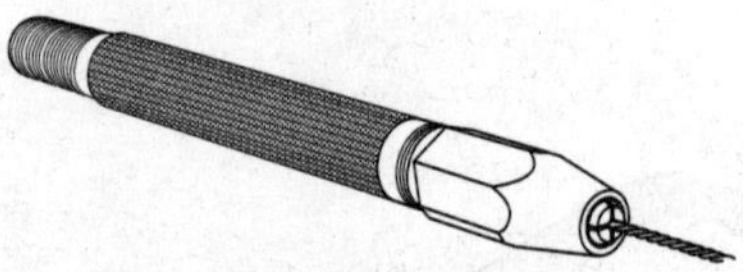

Pin Vise

Pin Vise: Used for twisting wire. It is adjustable to all sizes of wire. The wire is passed through the vise. Tighten the vise on the wire tight enough to still slide it. Make sure the end of the wire nearest the head of the vise is held securely. Tug gently on the vise as you twist the wire. Take no more than 1/4" - 1/2" bites. The number of twists will determine if you have a fine or coarse pattern. **IT IS IMPORTANT TO REMEMBER THAT TOO MANY TWISTS CAN BREAK THE WIRE.**

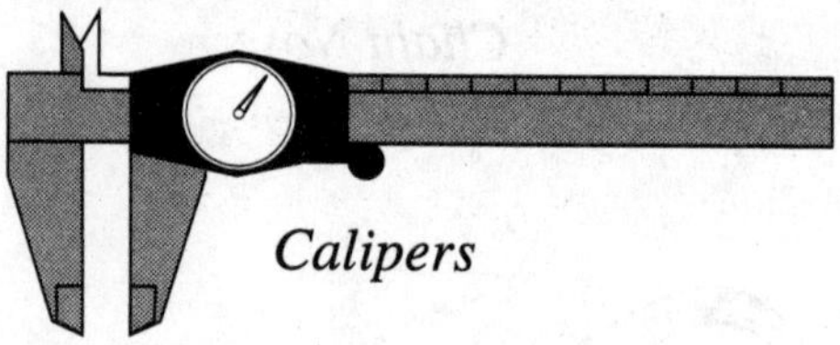

Calipers

Calipers: Used to measure the size of stones for rings (so wrap wires can be accurately placed).

Jeweler's Vise

Jeweler's Vise: A very small bench vise 4" or smaller is used in making bracelets and other jobs you will discover.

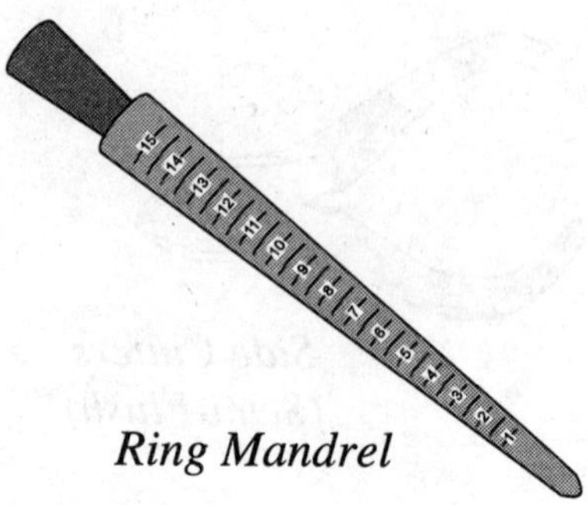

Ring Mandrel

Ring Mandrel: (Grooved or ungrooved) Used for sizing rings.

18" Steel Ruler: (Corked back) Used for measuring wire and making other measurements.

Small Pocket Knife: (With pointed blade) Point is used to create small openings for wire to be forced through.

Jeweler's File: For filing down rough edges of cut wire.

Opti-Visor: For magnification — especially when making free form rings.

Ring Sizer Set: For measuring fingers. **MAKE SURE THE RING SIZES ARE CALIBRATED WITH THE RING MANDREL.**

Plastic or Leather Hammer: Used for shaping rings on the mandrel and making bracelets.

Compartmented Box: (Wood, plastic, or metal) For storing odds and ends. Should be at least 12" x 12" with eight to ten compartments.

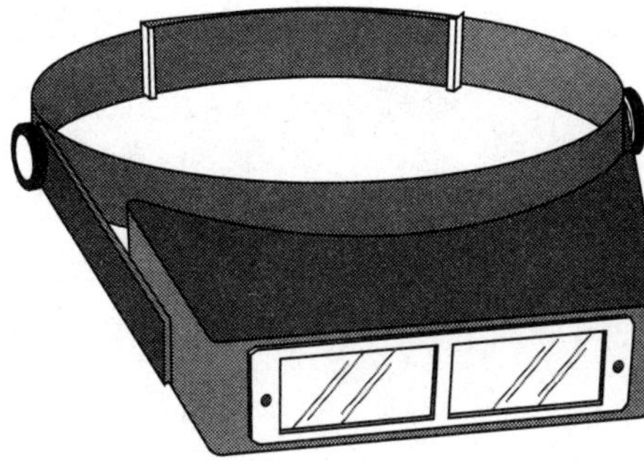

Opti-Visor

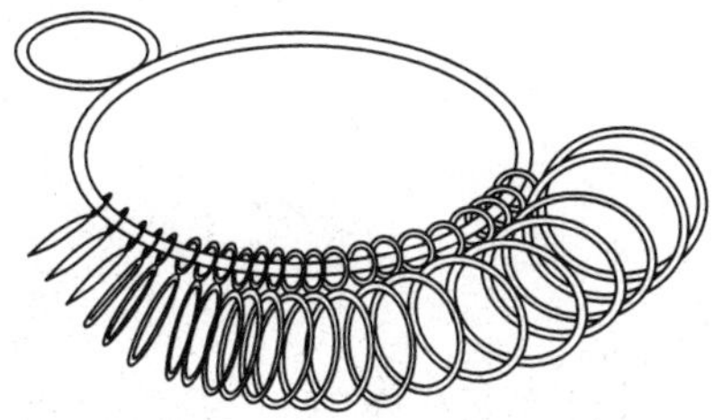

Ring Sizer Set

Rawhide Mallet

#1 WIRE

A. Wire for Finished Jewelry

Brass Wire

Round, non-tarnishable. Can be purchased at your local lapidary, craft, or hardware store.

20 gauge (.032)

22 gauge (.025)

24 gauge (.020)

Square, (practice wire) Can be purchased at your local lapidary store or from jewelry related catalogs.

Copper Wire

Round

18 gauge (.040)

20 gauge (.032)

22 gauge (.025)

24 gauge (.020)

Square

21 gauge (.028)

Can be purchased at your local lapidary store or from jewelry related catalogs.

B. Inexpensive Wire

Local hardware stores and electrical supply stores are a "treasure house" of all different types and sizes of good inexpensive wire -- everything from copper, zinc, to aluminum. Be brave. **EXPERIMENT**!!!

#2 STONES

A. Glass Beads: 4mm to 12mm. A wide variety of other bead material is available at your local craft store.

B. Glass/Acrylic Cabochons: From 10 x 12mm to 30 x 40mm. These can be smooth or faceted (looking).

C. Any baroque shaped object, not larger than a fifty cent piece, and arrowheads. Beach glass looks good in brass or copper wire.

#3 MISCELLANEOUS

A. Jeweler's cloth

B. Piece of felt 12" x 18"

C. Small work table

D. Two small plastic boxes with lids (for scrap wire)

E. Fine line felt tip pen

F. Small roll of paper tape

G. Cotton bags (for wire)

H. A small carrying case (for tools and wire)

#1 SINGLE WIRE PENDANT

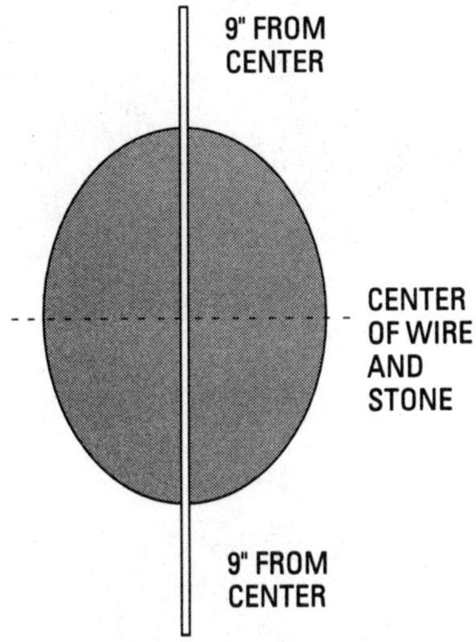

Diagram 1

A. Select a 30 x 40mm glass (or acrylic) oval cabochon.

B. Cut a piece of round, 20 gauge brass wire, 18" long if you want only one frame around the stone. For two wires around the stone, cut it 25" long. For a triple frame, cut it 32" long. (See paragraph P, pages 19 and 20.)

C. Place the center of the wire on the back of the cabochon and sandwich it between your thumb and forefinger of the left hand to hold it firmly. *Diagram 1*

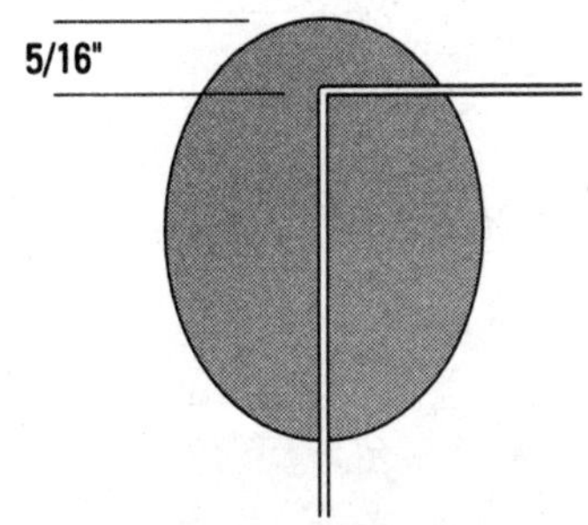

Diagram 2

D. Make sure there are equal lengths of wire extending beyond both ends of the cabochon. *Diagram 1*

E. Using the flat nose pliers, bend the the wire at a right angle (90 degrees). This bend should be 5/16" below the outer edge at the top of the stone. *Diagram 2*

F. Holding the wire flat up against the cabochon, press hard with the left thumb and at the same time use the left forefinger to wrap the wire around the top front face of the stone. Keep wrapping until it is across the back of the cabochon. *Diagram 3*

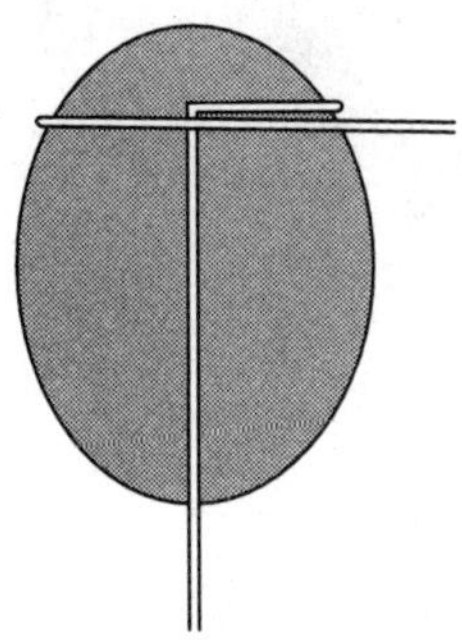

Diagram 3

G. Remove the cabochon, and while holding the loop in place with the flat nose pliers (very tightly as indicated in *Diagram 4*), wrap the wire that goes around the cabochon one and a half times around the wire that runs the full length of the cabochon.

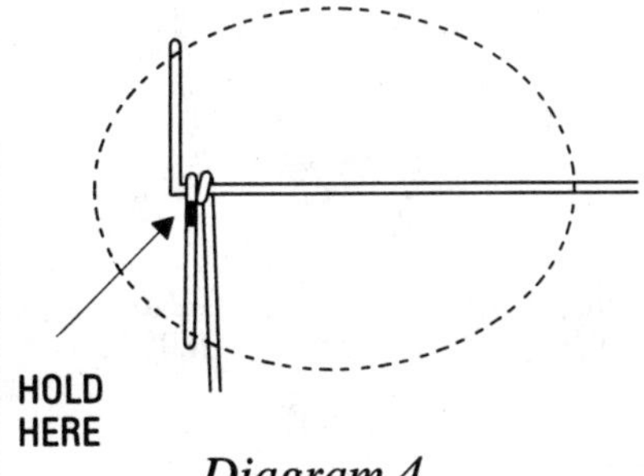

Diagram 4

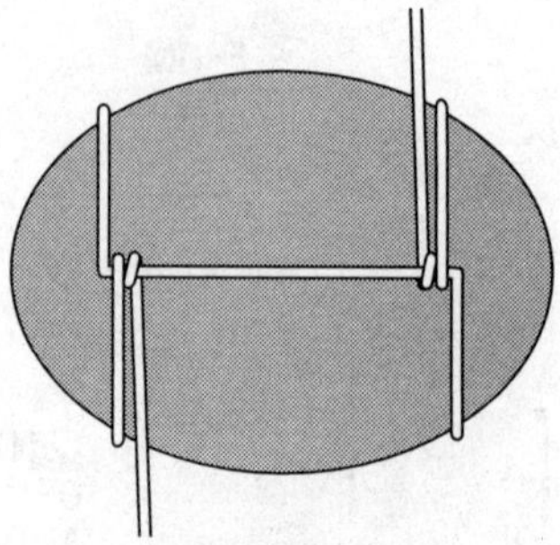

Diagram 5

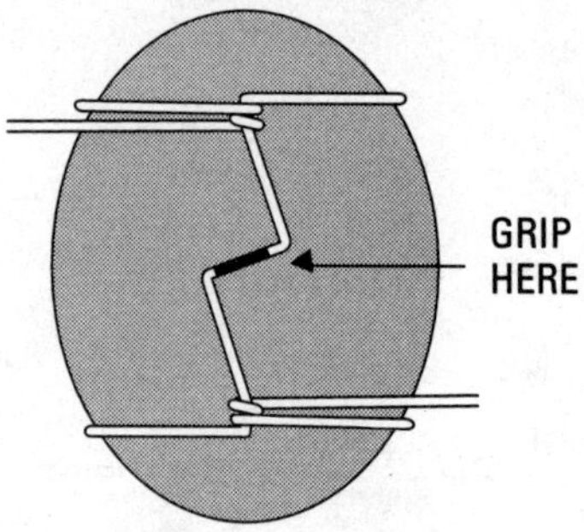

Diagram 6

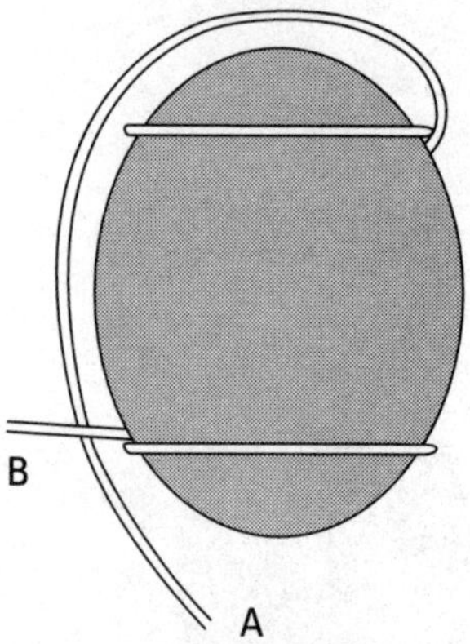

Diagram 7

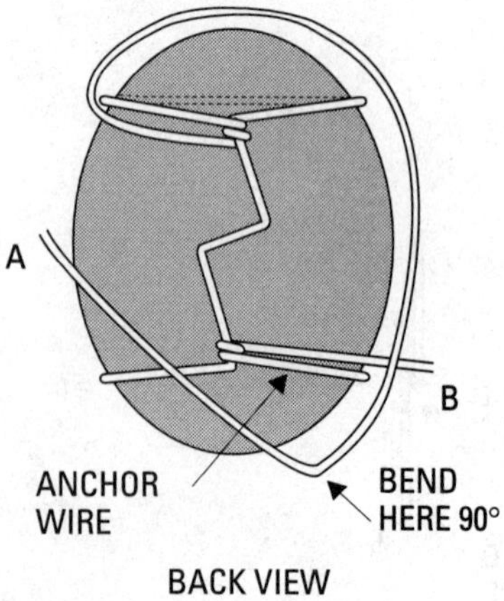

Diagram 8

H. Reinsert the same end of the cabochon in the completed loop making sure the back, front, top, and bottom is as it was before the loop was made.

I. Repeat steps F, G, and H for the other end of the cabochon. *Diagram 5*

J. After the cabochon has been reinserted into both loops, grip the wire that runs the length of the back of the cabochon with the flat nose pliers and turn clockwise about 1/4 turn. Do it **slowly,** being careful not to break the wire. This will tighten the loops around the end of the cabochon and lock the wires securely. *Diagram 6*

K. Hold the entire assembly in your right hand so that the front is facing you. With the left hand, very gently bend wire A in an arc to the left so it will follow the contour of the left outer edge of the cabochon. There should be no more than 1/4" of space between the wire and the edge of the cabochon (the distance will be greater at the top). *Diagram 7* Please note the position of wire A and wire B at this time.

L. With the back of the assembly facing you, (holding it in your left hand), put a 90 degree bend in wire A as indicated. (Use the right thumb and forefinger.) *Diagram 8* Bend it just beyond what is to be called the anchor wire. *Diagram 8*

M. Now insert wire A as indicated (*Diagram 9*) and pull it gently until it reaches the 90 degree bend. **DO NOT PULL IT ANY MORE.**

N. Bend wire A after it has been inserted and pulled properly as indicated by the arrows. This can be done by gripping the wire on the end with the flat nose pliers and tugging gently to make it snug. Use the chain nose pliers to make it tight. *Diagrams 10 & 11*

O. With wire B, repeat steps K, L, M and N.

P. Put a jump ring on one end of the assembly. Put a chain through it, and now you have a cabochon framed in a single wire for an attractive pendant. For a double wire frame use a 25" piece of wire--**do not** cut it off as in *Diagram 11.* Just tighten it and repeat steps K, L, M and N.

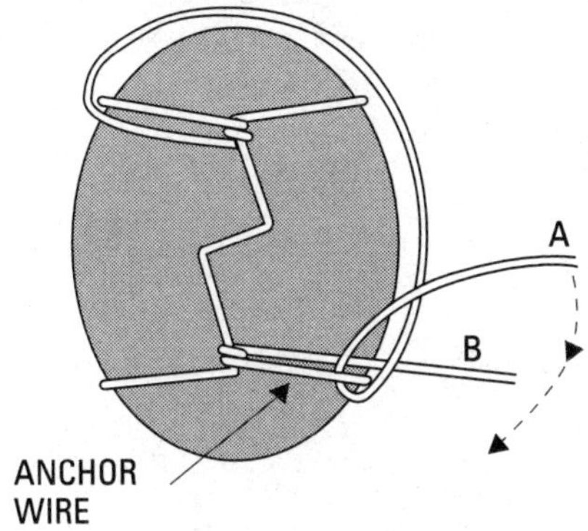

Diagram 9

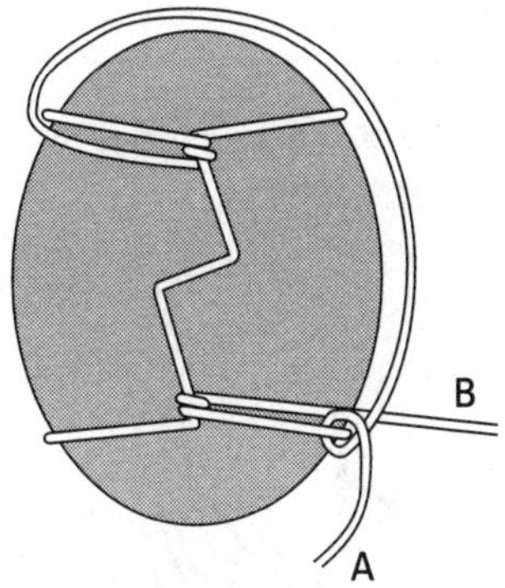

Diagram 10

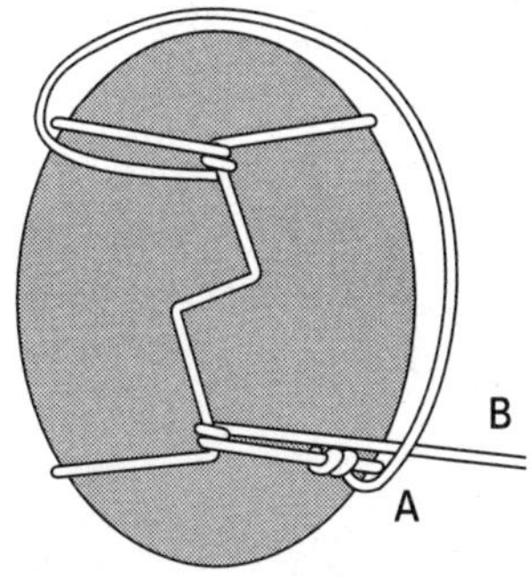

Diagram 11

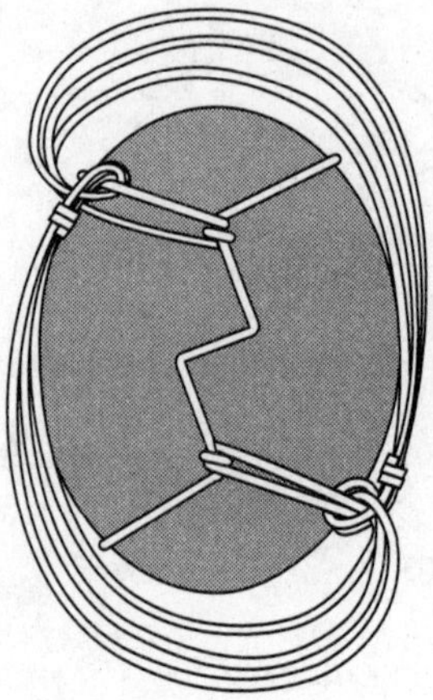

Diagram 12

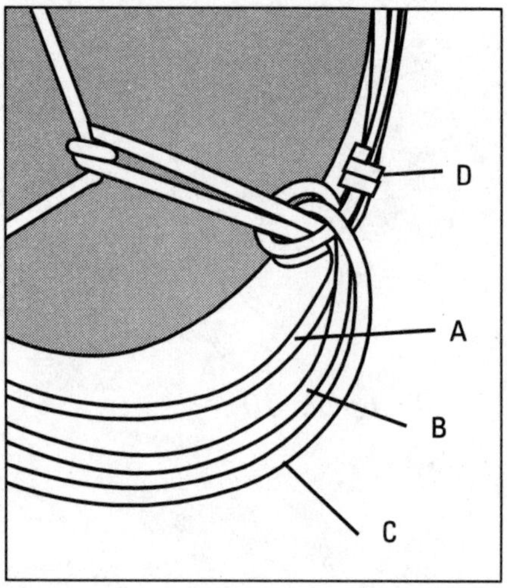

Diagram 12A

Do likewise, with a triple wire frame and a 32" piece of wire. The tie-down is different when there is more than one frame wire. There is enough wire left to wrap it securely around all of the other frame wires.
Diagram 12

In reality, wire A in *Diagram 12A* cannot be seen when finished. It is under B and C.

Likewise, the same is true for wire B because it is under wire C.

Wire C is on top and is extended to wrap around all three wires at Point D, 2 1/2 times.

Q. Other suggestions for this wrap: For any cabochon 10 x 14mm to 18 x 25mm, a single wire frame is sufficient. For cabochons larger than 18 x 25mm or 30 x 40mm, 2 or 3 frame wires would be more appropriate.

#1 NARROW OVAL PIN

A. Select a glass (or acrylic) cabochon 20 x 40mm and cut a piece of brass wire, 20 gauge, non-tarnishable 16" long.

B. This is a variation of the single wire pendant. There is a very subtle change in procedure at the beginning. In the previous chapter note *Diagram 5* how each of the wires comes out from the stone in opposite directions. For this pin, the wires should be coming out from the stone in the **SAME** direction. *Diagram 13*

C. This can be done as follows: Repeat step F (*Diagrams 2 & 3*) page 17. Instead of continuing the wrap as indicated in paragraph G, put the wire that goes around the stone **UNDER** the wire that goes down the back of the stone, then wrap it counterclockwise 2 times around so the wire exits to your right as in *Diagram 13.*

D. Reinsert the stone in the completed loop and complete the wrap at the other end of the stone as instructed in paragraph I, page 18.

E. For completion of this project, follow this instructional diagram. It is self-explanatory. Please note where the tie-downs are. *Diagram 14*

F. Make sure the tie-downs are tight for added stability.

G. When the pin is attached, make sure the entire pin is well behind the stone and hidden from view from the front. *Diagram 15* (For a refresher on how to make pins, see Volume 1, page 100.)

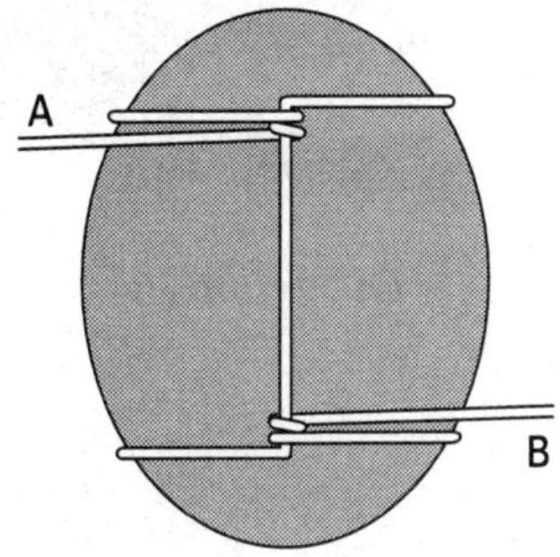

Diagram 5, Chapter 3

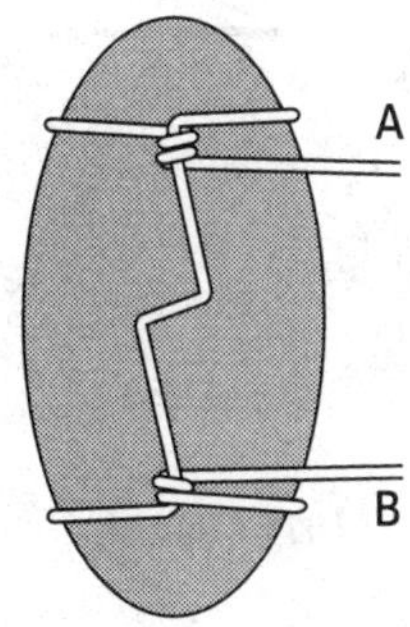

Diagram 13

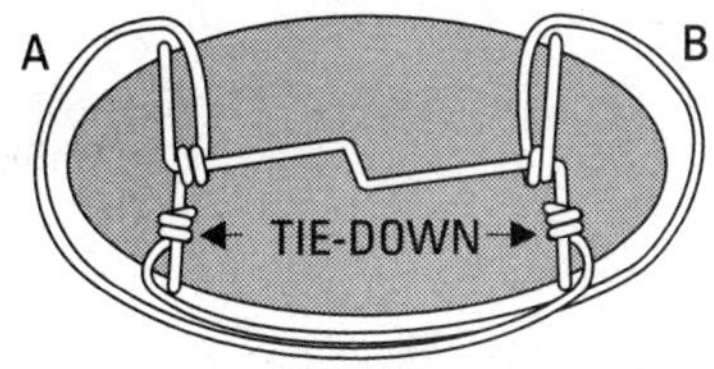

Diagram 14

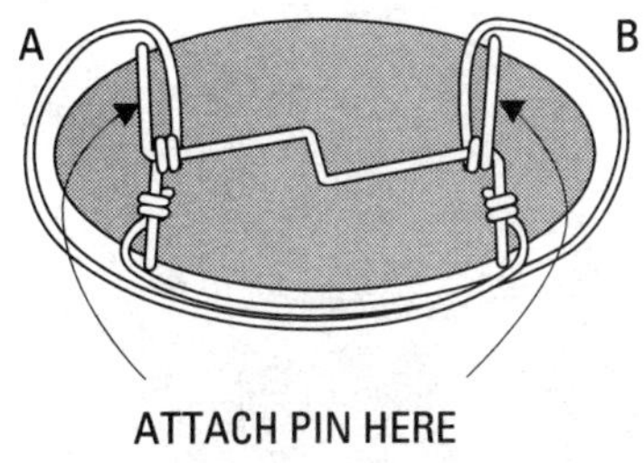

Diagram 15

Diagram 16

H. A popular variation of this piece is as follows: After bending and shaping wire A and wire B, put one small loop in one of them and then put two small loops in the other, so there is a total of 3 loops evenly across the bottom before they are tied down. From these loops can be hung matching drilled stones or beads. *Diagram 16*

#2 PENDANT

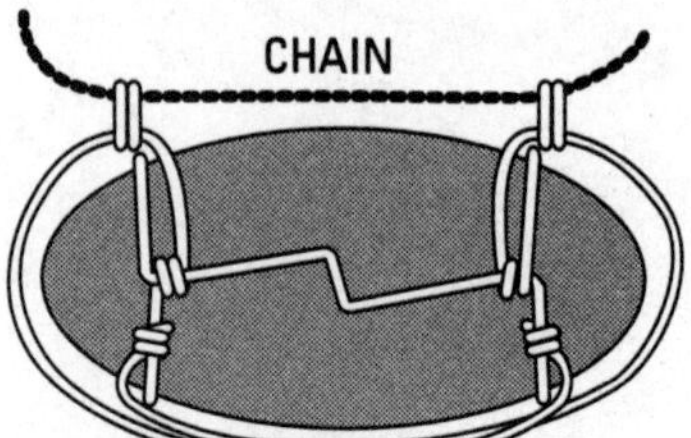

Diagram 17

A. One more variation of this is also popular: If you wish a pendant, then attach jump rings as indicated to wire A and wire B (make them large enough to run a chain through). *Diagram 17*

#3 EARRINGS

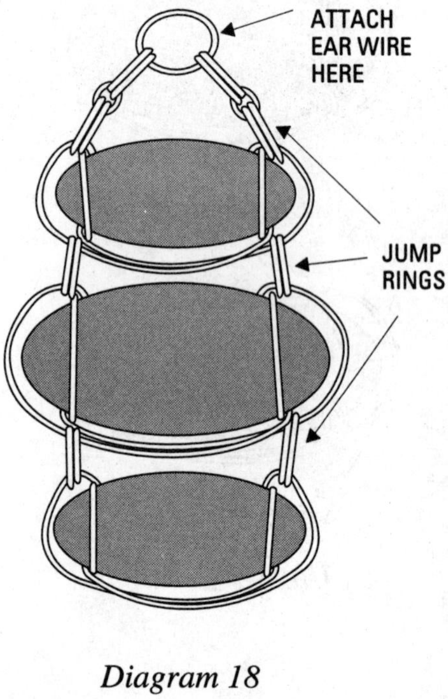

Diagram 18

A. Another variation (which is most effective on long narrow ovals ranging in size from 16 x 8mm to 25 x l0mm) is to attach three of these together in parallel by interlocking them with jump rings. All three of them can be of the same size, or they can be a series of three stones of graduated sizes from small to large.

It is suggested that larger stones be used for the pendant, and smaller stones be used for the earrings. A total of 9 stones is needed for this project. Length of wire will vary with the size of the stone. Earrings can be graduated sizes of stones. Also note the finished earring diagram, and how it is attached to the ear wire. *Diagram 18*

#1 MULTI-GLASS BEAD PENDANT

There is a lot of room for flexibility and creativity in making this pendant. There is a multitude of options. This chapter deals with only the option that produced (in my opinion) the most outstanding results. Have fun and go for it !

Materials: One piece of 22 gauge or 24 gauge wire at least 33" long,

13 drilled glass beads 6mm, 7mm, or 8mm in size. They can be all the same color or otherwise.

A. Bend one end of the wire up about three inches (this will keep the beads from sliding off).

B. Insert the other end of the wire through all thirteen beads. *Diagram 19*

C. After all of the beads are on the wire, form a circle and gently tie an overhand knot with both of the wires. Wrap the short piece around the wire going through the bead and snug it up so the tie will not come apart. You can make it tight later. After the tie, make sure there is an open space between the first and last bead of about 1/2". *Diagram 20*

D. **DETAIL OF TIE-DOWN**: Wrap wire A twice around wire B. Make it snug but not permanently tight. **(AS EACH BEAD IS WRAPPED AND SECURED, THE REST OF THE SPACE BETWEEN THE BEADS WILL GRADUALLY GROW SMALLER AS THIS SPACE BECOMES OCCUPIED WITH WIRE. EVENTUALLY MOST OF THE SPACE WILL BE USED.)**

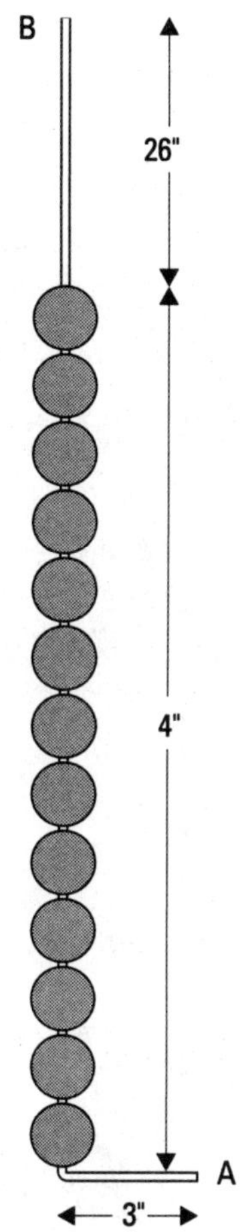

Diagram 19

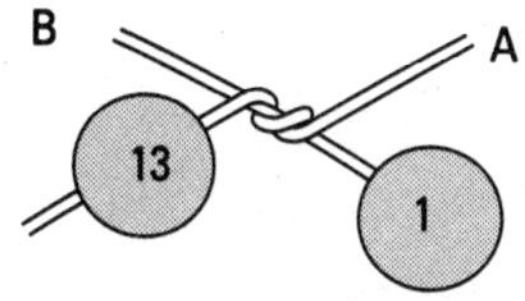

Diagram 20

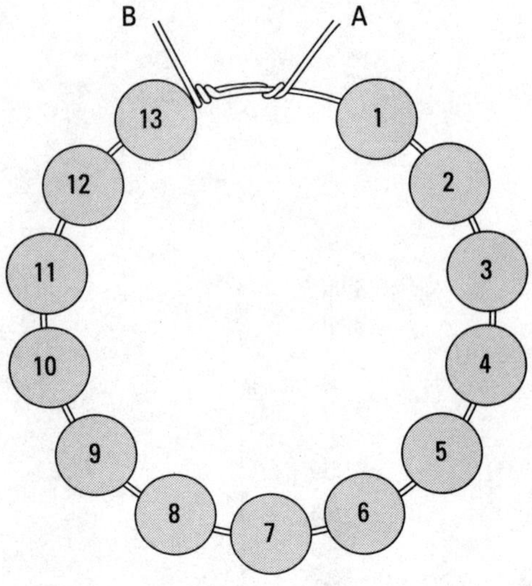

Diagram 21

E. Proceed as follows: Push bead #13 up to the tie-down and wrap wire B one and one half times around the wire that goes through it. (Wire A) The thirteen beads should now resemble a distorted circle with wire B perpendicular to the outside circumference. *Diagram 21*

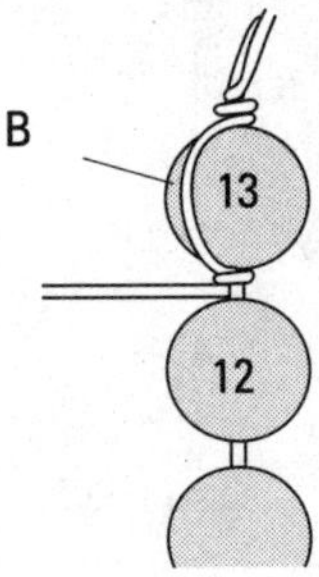

Diagram 22

F. Bring wire across the outer surface of the bead diagonally and wrap it around the wire between bead #13 and #12 one and one half times. (*Diagram 22* enlarged for clarity.)

G. Repeat this process all the way around to and including bead #1. Minor adjustments are made continually as the circle forms. Squeeze the assembly gently to maintain a symmetrical 360 degree circle, and check to make sure that each wire is **DIAGONAL ACROSS EACH BEAD**.

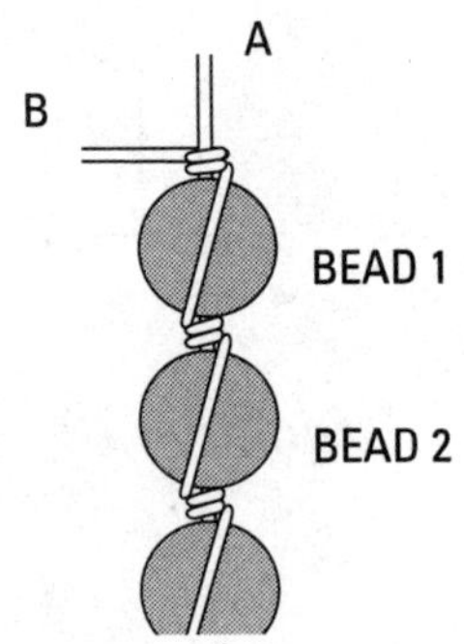

Diagram 23

H. When bead #1 is reached and completed like the other 12, proceed as follows: *Diagram 23*

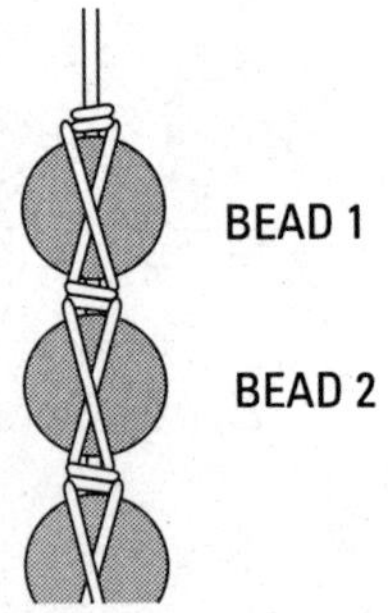

Diagram 24

I. After the diagonal across bead # 1, wrap wire B around wire A one and one half times and bring it back across the other diagonal on bead #1 so an "X" is formed on the outer surface of bead #l. *Diagram 24*

J. Repeat this process for beads #2 - #13. Be sure to wrap the wire 1 1/2 times around the wire in-between the beads. It will be necessary to wrap it over on top of the wire that is already there. By this time, the circle of beads should be a very stable 360 degree circle with wires A and B next to and parallel to each other and perpendicular to the circle. *Diagram 25*

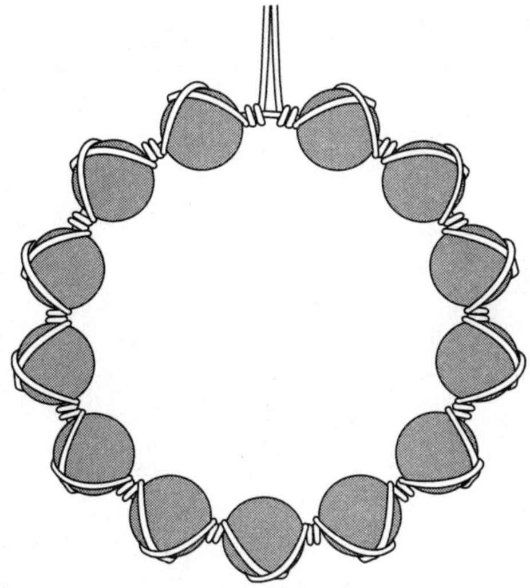

Diagram 25

K. Twist wires A and B together very tightly. Make sure they are about 1 1/2" long after they are twisted together. *Diagram 26*

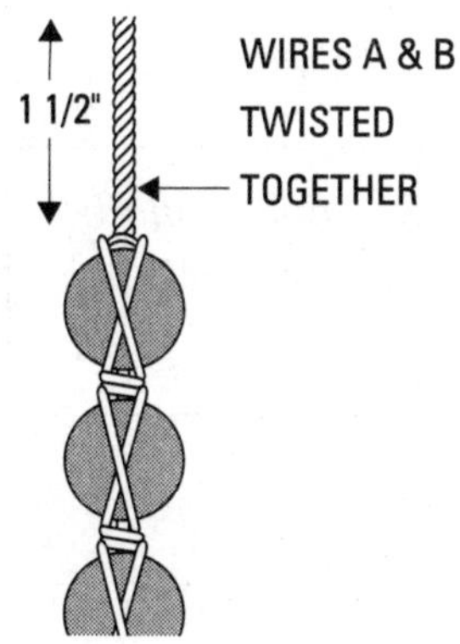

Diagram 26

L. Leaving this stem of 1 1/2" is a matter of choice, depending on what kind of loop you want. You can have a single, double, or triple loop from which to hang it. The diagram shows a single loop which takes less than 1 1/2". *Diagram 27*

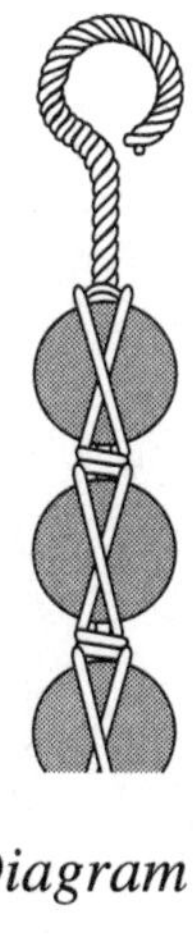

Diagram 27

SUMMARY: VERY IMPORTANT

See color photo on cover or insert for finished piece

1. You can use three or five beads to make a set of matching earrings.

2. This pendant looks best when done with eleven or thirteen beads, 6mm, 7mm, or 8mm in size, and a single piece of 24 gauge or 22 gauge round. Using this thin wire is difficult, but **IT CAN BE DONE**. Just be careful not to break it.

3. On your first attempt, use eleven 12mm beads and a 30" length of round 21 or 20 gauge brass wire.

4. If you don't leave any space between the beads at the beginning, tie the wire in a double overhand knot leaving 1/2" between the first and last bead. Then all you have to do is to bring it diagonally across each bead and then pull it up and wedge it permanently between the beads. The end product looks good because each bead has its own "arch". **TRY IT!! YOU'll LIKE IT.**

5. If you have a problem following the steps about how to begin or end this project, don't fret--experiment. It's exciting to explore the unknown! You may even find a better way to do it.

6. This pendant can also be made using non-round beads mixed with round beads. Remember, the only limit is your imagination. Don't ever say "I can't"-- Say "I'll try!!"

7. Every multi-bead pendant you make will somehow look different. Your own personal touch will lead you to modifications never before thought of by anyone! Such is the joy of wirewrapping!!

#1 CHARM BRACELETS AND TENNIS BRACELETS

Preliminaries for this project have been brought forth in Chapter 4, Volume 1, *MOODS IN WIRE,* pages 49 through 53. This involves the three wire braided bracelet. There is no need to repeat those instructional pages here. What follows is simply another idea added to the original idea. The three wire braided bracelet must be completed first in order to complete the instructions in this chapter.

1. Complete the three wire braided bracelet as instructed in Volume 1, pages 49 through 53.
2. Select seven 6mm drilled glass beads or any odd-shaped drilled glass pieces.
3. Cut a length of 20 gauge round brass wire 12 to 15". This wire will be used as needed to make the jump rings (one by one). If this is used up, another piece can be cut as needed.
4. Using the round nose pliers, proceed as follows: *Diagram 28*

A. Grip the end of the wire in the jaws of the round nose pliers at the larger end of the jaws. *Diagram 29*

B. Then push the wire downward until it is all the way around the lower jaw of the pliers. **(DO NOT ROTATE THE PLIERS. PUSH THE WIRE AROUND THEM AS SHOWN.)** *Diagram 30*

C. Remove the jaws of the pliers from the partially completed circle of wire.

D. Insert this partially completed circle through an opening on one edge of the braided bracelet as shown in *Diagram 31.*

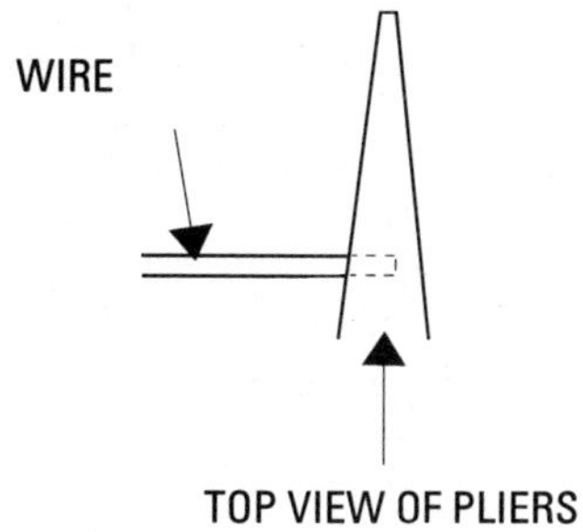

Diagram 28

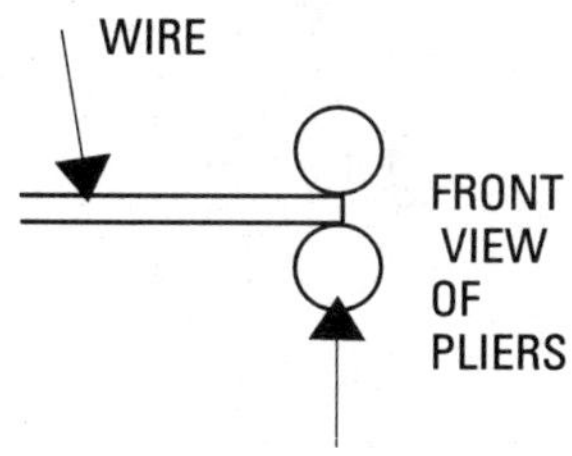

Diagram 29

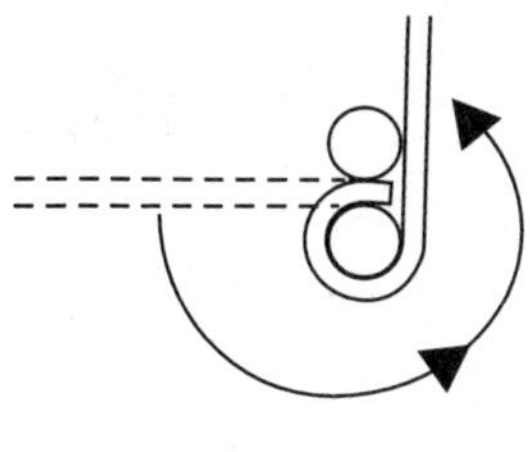

Diagram 30

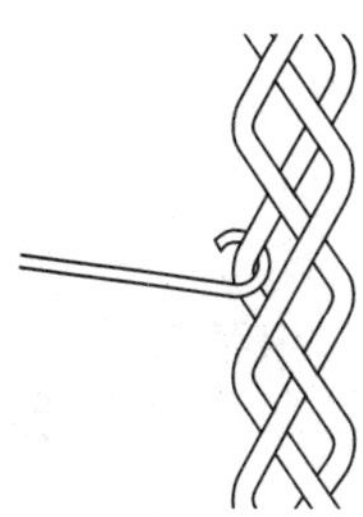

Diagram 31

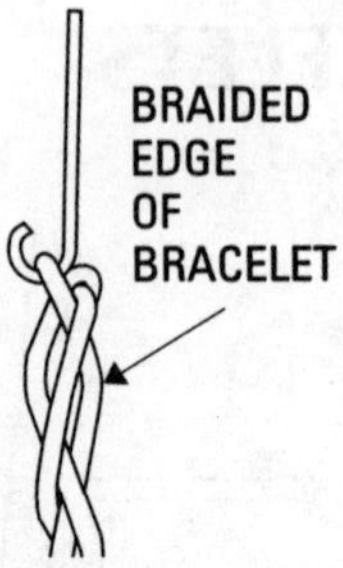

Diagram 32

E. After hooking the outer edge, let the bracelet dangle (like a fish on a hook). Make sure as it hangs, the long end of the wire is on your right, and the open part of the "hook" is on your left. *Diagram 32*

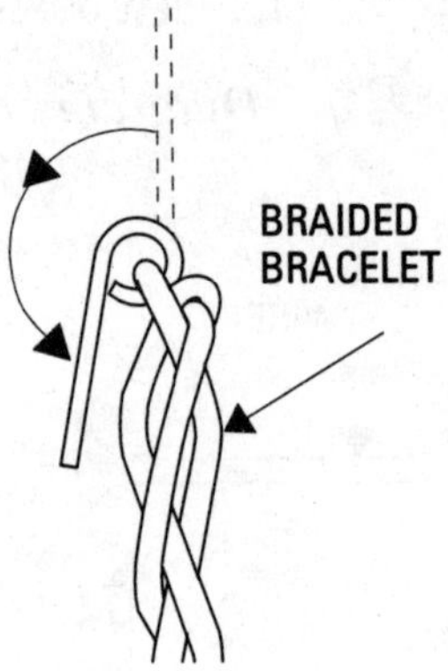

Diagram 33

F. Reinsert the bottom jaw of the round nose pliers in the partially completed circle of wire, making sure you are using the thicker part of the jaws. With a firm grip, hold the pliers and push the wire to your left and down. This will close the circle and continue the progress of the jump ring. *Diagram 33*
(DO NOT ROTATE THE PLIERS. PUSH THE WIRE AROUND THEM.)

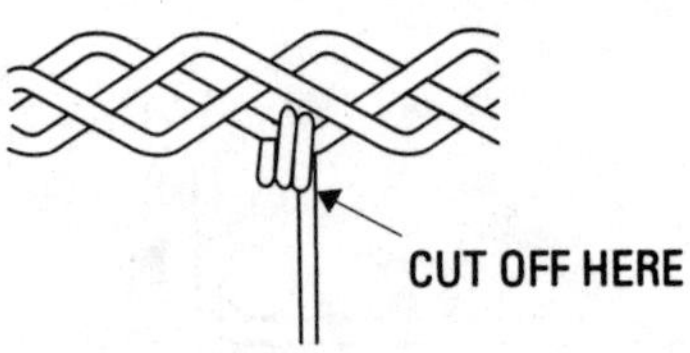

Diagram 34

G. Ease up on the pliers grip and move the long wire (of the jump ring) to the right while still on the jaw of the pliers. Grip again firmly and continue pushing to the left and downward until the other end of the jump ring shows again. Cut it off. *Diagram 34*

H. Repeat this process six more times so that they are evenly spaced around the **SAME EDGE** of the three wire braided bracelet.

Suggestions for spacing beads evenly:

#1 Put directly opposite the clasp.
#2 Put at left center.
#3 Put at right center.
#4 Put at upper right diagonal.
#5 Put at lower left diagonal.
#6 Put at upper left diagonal.
#7 Put at lower right diagonal.
Refer to *Diagram 35* for clarity.

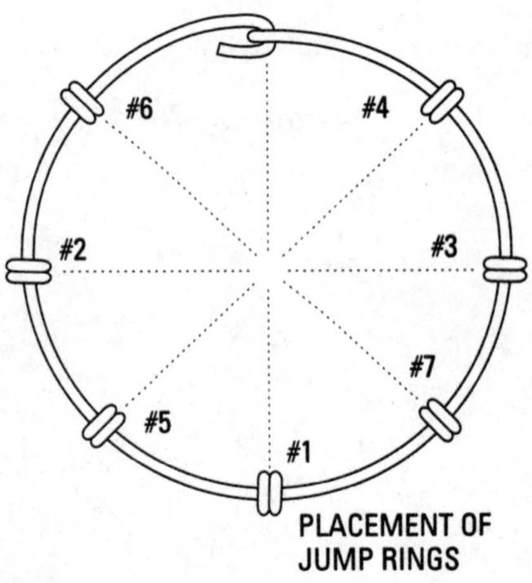

Diagram 35

I. Each bead is to hang freely from each of the seven jump rings. (You can hang the beads directly from the braided bracelet, but there is not a very good "dangle" effect. Use jump rings to get the maximum effect.) For those of you with experience in wirewrapping, you may want to use head pins to complete this next phase. It will save time and labor, otherwise continue as follows:

J. Cut a length of 22 gauge round brass wire 10 to 14" long.

- On one end of this wire make a small hook with the round nose pliers. *Diagram 36*
- Clip off a small portion of the hook as indicated in *Diagram 37*
- Then using the flat nose pliers, squeeze it together. *Diagram 38*
- This will prevent the bead from slipping off. (The same thing is accomplished by using a headpin.)

K. Put a bead on the wire and slide it down to the end just completed. *Diagram 39*

L. From the top of the bead, measure 1" and cut it there. *Diagram 40*

M. Using the small end of the round nose pliers, make a hook and put it through one end of the jump rings. *Diagram 41*

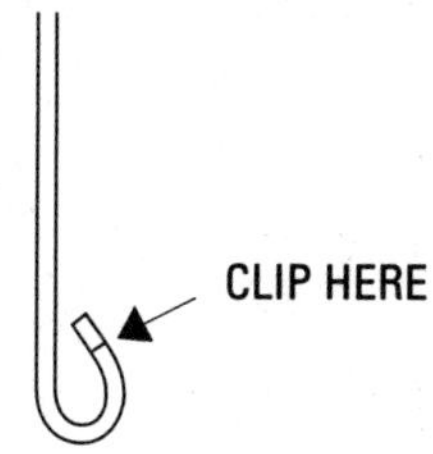

Diagram 36

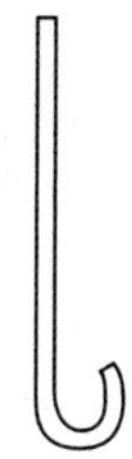

Diagram 37

Diagram 38

Diagram 39

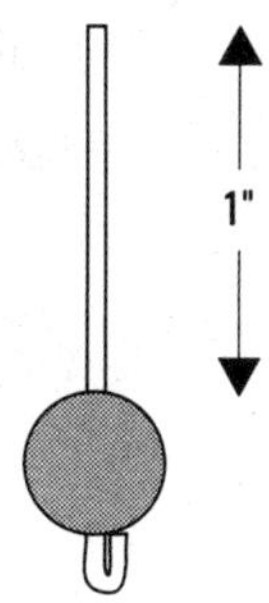

Diagram 40

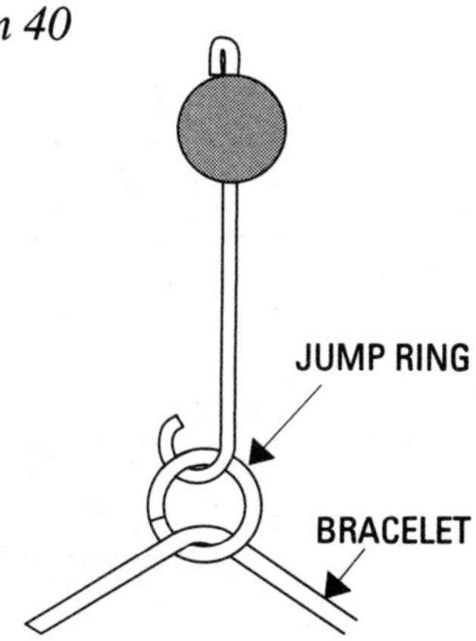

Diagram 41

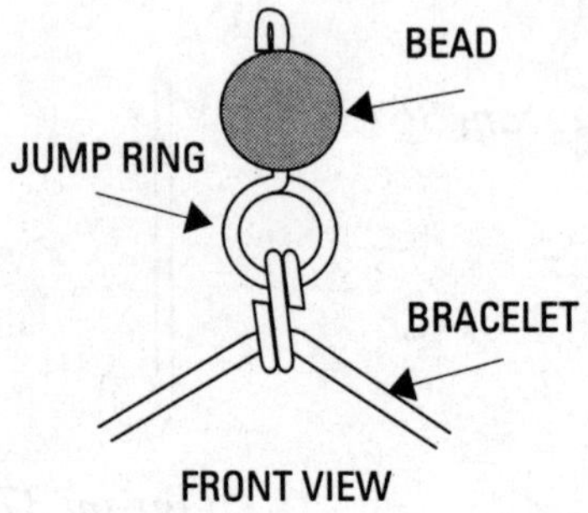

Diagram 42

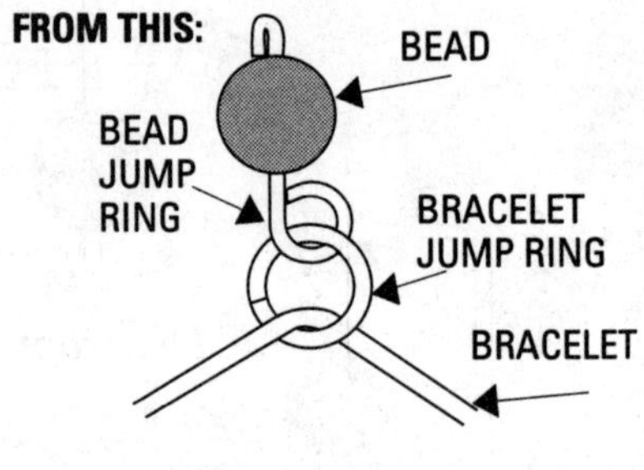

Diagram 43

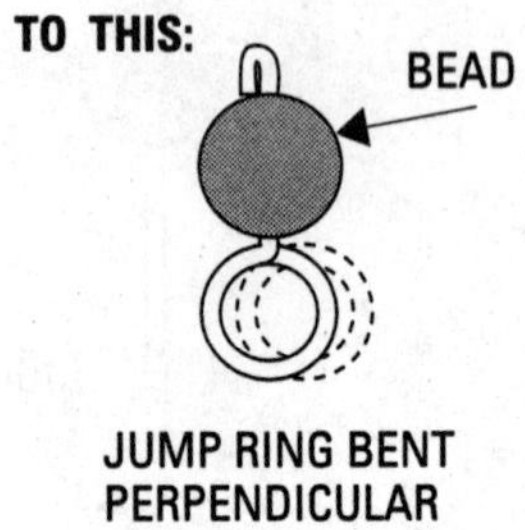

Diagram 43A

N. Using the same method as putting the jump rings on the braided bracelet, push it around the round nose pliers so the circle of wire is up against the bead. *Diagram 42*

O. With the small end of the round nose pliers, tilt the loop back so the stem of the bead jump ring is perpendicular to the bead. *Diagrams 43 & 43A*

P. Repeat this bead attachment process for the other six beads. The bracelet is now complete.

Summary: Very Important

1. There is, literally, no end to the variations you can create with this "charm" bracelet. Different color combinations are popular; different sizes of beads ranging from multi-face glass beads, to rondelles, to V-shaped and tear drop, or even cloisonné beads can be used.

2. These bracelets can also be interspersed with inexpensive pewter figures. Tiny little glass stars or hearts are very popular with children. Children's bracelets should be less "cluttered" than adult bracelets. Also make them 5 to 6" long from hook to ring (fastener) instead of the normal 7" adult size.

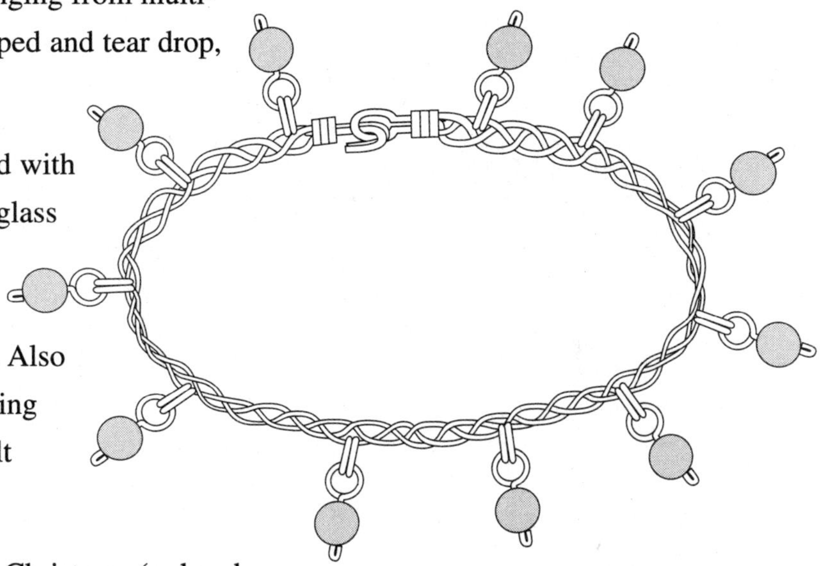

Finished Charm Bracelet

3. You can also make theme bracelets for Christmas (red and green), or Halloween (orange and black), or red, white, and blue for patriotism. The number of charms that can be put on a bracelet is strictly up to you. I have done as few as three (triplets) to as many as thirty-six (birthday). Matching earrings would be simply a miniature version of what has gone before.

4. A final variation of this bracelet is already on the commercial scene and has been very popular. I am talking about the "tennis" bracelet for the ankle. Just eliminate the three wire braided part and put together a series of at least thirty five jump rings (depending on how large or small your ankle is). Also, I challenge you to come up with a good fastener. It really isn't that difficult. Good luck and have fun!!

See color photo on cover or insert for finished charm bracelet.

CHAPTER 7 — SIMPLE EARRINGS

GRANT HODGES

#1 SIMPLE SINGLE WIRE EARRINGS

In the world of wirewrapping, without a doubt, making earrings has a great, if not the greatest potential, for expressing your own personal style. Earrings not only make a statement about their creator, but also about the wearer as well. When a customer at my table suddenly says, "Hey ! I really like those earrings!!", I feel the silent warm glow of joy in knowing I have communicated with a stranger without uttering a single word. You too can experience the reward of having your mind on the same frequency as someone else!

There is a wide variety of materials available for this particular medium. It's an open invitation to be different. The possibilities are endless. Find your own style and personal expression. Try to avoid the obvious, but if you can't, go to it, and by that time you will have thought of many variations, some of which will be "cool." **HAVE FUN !!!**

A. You will need two drilled, round glass beads, 6mm or 8mm in size, and a piece of round or square brass wire 4 to 6" long. If you use square wire, then twist it.

B. Insert the wire into the bead until the end of it sticks out 1/4". *Diagram 44*

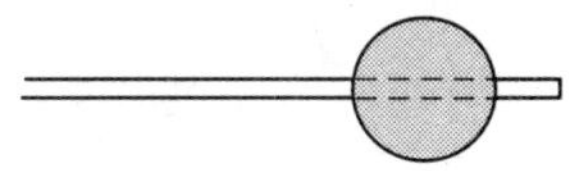

Diagram 44

C. Grasp the short end of the wire with the flat nose pliers so the pliers are perpendicular to the bead (not parallel). *Diagram 45*

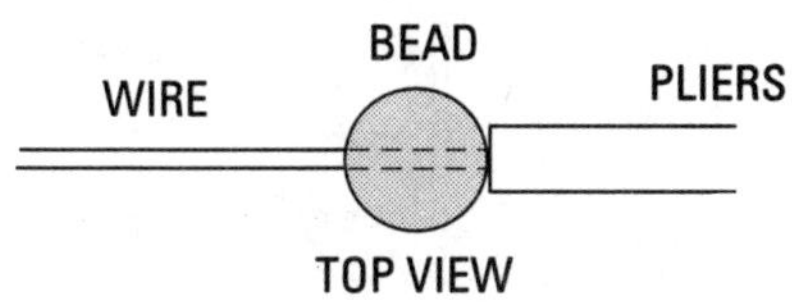

Diagram 45

D. Holding that end firmly, bend the long wire up and around the bead back toward the pliers. *Diagram 46*

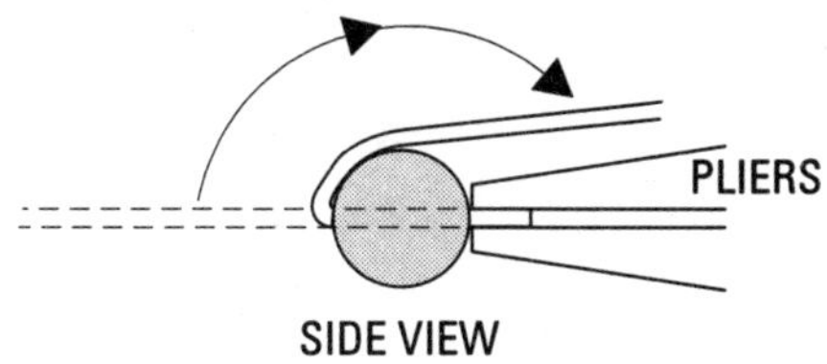

Diagram 46

E. Shift position of the pliers to where the longest part of the wire comes out of the bead. *Diagram 47*

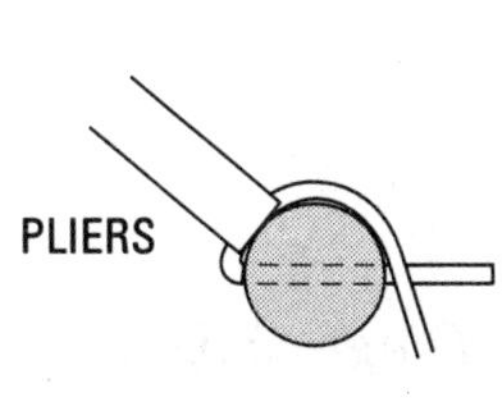

Diagram 47

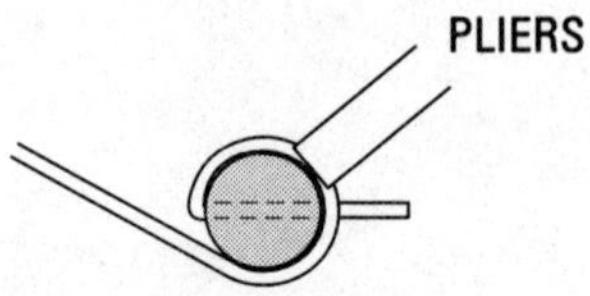

Diagram 48

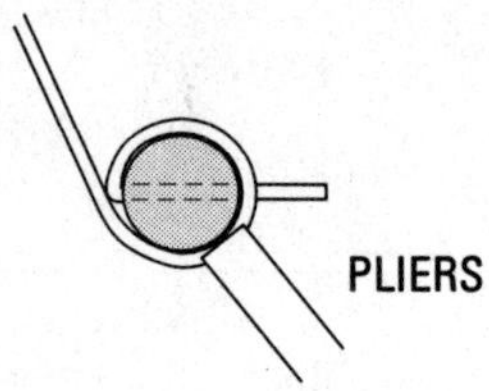

Diagram 49

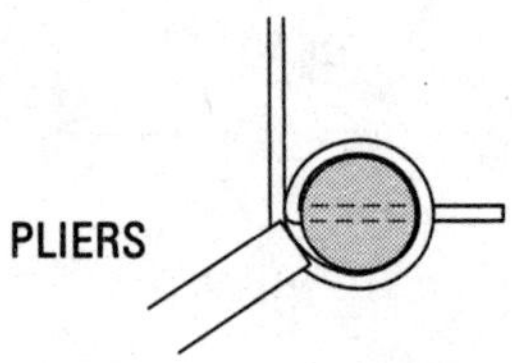

Diagram 50

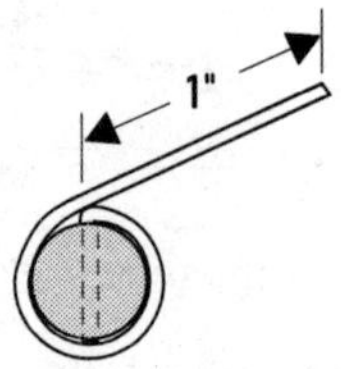

Diagram 51

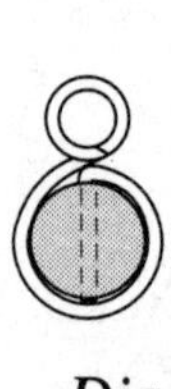
Diagram 52

F. Continue bending the wire around the bead as tight as possible, at the same time shifting the pliers so they grip the wire close to where the bending is. *Diagrams 48, 49, & 50*

G. Cut off the short wire right at the surface of the bead and then nudge the wire (going around the bead) over the cut end to hide it.

H. Cut off the longer wire about 1" beyond where it meets the wire coming out of the bead. *Diagram 51*

I. Using the round nose pliers, grip this wire on the smaller end so that the pliers are perpendicular to it, and roll it backwards into a completed round loop (for an ear wire to be attached). See *Diagram 52* for this completed project.

J. The loop can be made smaller, depending upon your personal taste.

K. An interesting variation would be to make a loop halfway around the bead so other wrapped beads can be attached. *Diagram 53*

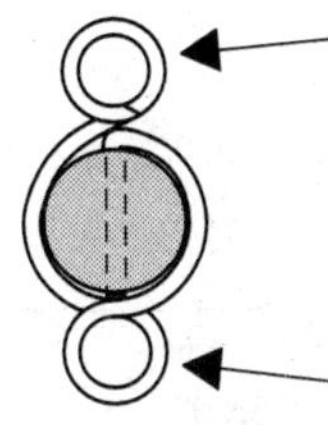

Diagram 53

SUMMARY:

This is a good basic, simple design with a minimum of materials and maximum flexibility for variety. A good project would be to make a 24" necklace using 20 gauge round wire and alternating 8mm ruby red glass beads with 8mm x 6mm long ovals. Do not connect them directly to each other. Use homemade jump rings with every connection. You can improvise your own clasp. Another variation on the earring you just finished is to make a long stem to which you attach the ear wire.

#2 SIMPLE SPIRAL EARRINGS

These earrings can be made very quickly and easily. You may use drilled or undrilled beads. The following directions are for **DRILLED** 12mm beads. Larger beads are easier to work with. As you become more acquainted with the system, you will be able to use smaller beads more easily.

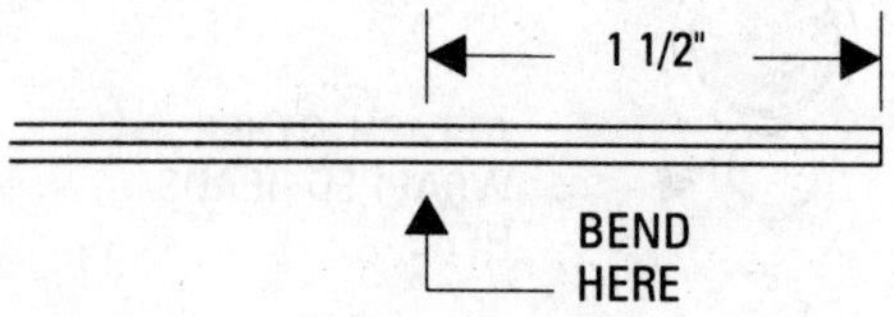

Diagram 54

A. Cut a piece of 20 gauge round brass wire about 6" long. Select two 12mm drilled beads.

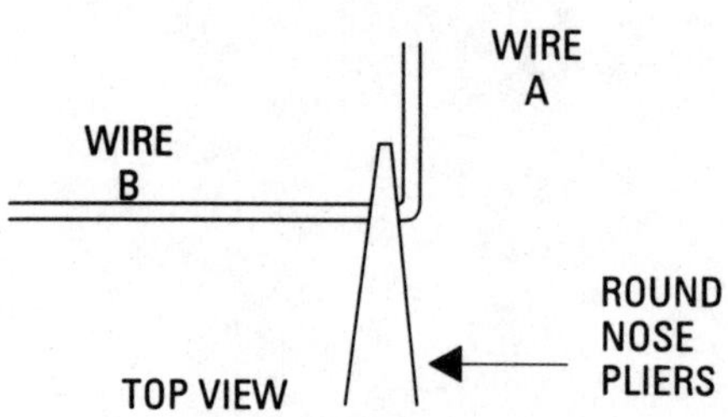

Diagram 55

B. Approximately 1 1/2" from one end of the wires, bend them up 90 degrees (bend them together to assure accurate duplication). *Diagram 54*

C. Grip wire B with the round nose pliers firmly as indicated in *Diagram 55.*

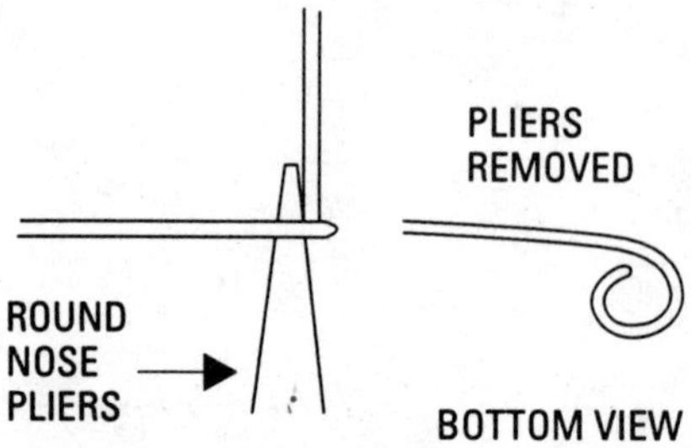

Diagram 56

D. Push wire B down and around the lower jaws of the pliers so it comes up the other side of the pliers. Withdraw the pliers from the loop, reinsert the top jaws immediately, and tighten your grip so you can continue bending the wire a full 360 degrees across the jaws of the pliers and also over wire A. *Diagram 56*

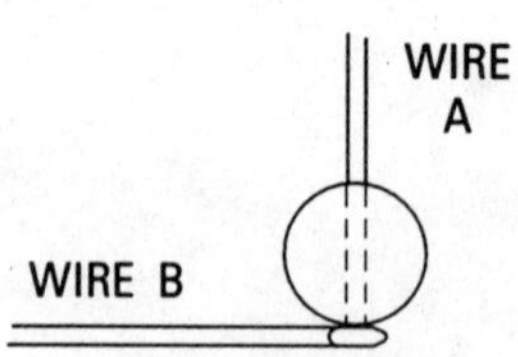

Diagram 57

E. This loop can be squeezed gently with the chain nose pliers. Insert wire A into the drilled bead. *Diagram 57*

GRANT HODGES

F. Grip wire A with the flat nose pliers very firmly just above the bead and proceed to wrap wire B in a spiral around the bead no more than 2 1/2 to 3 times around a 12mm bead. *Diagram 58*

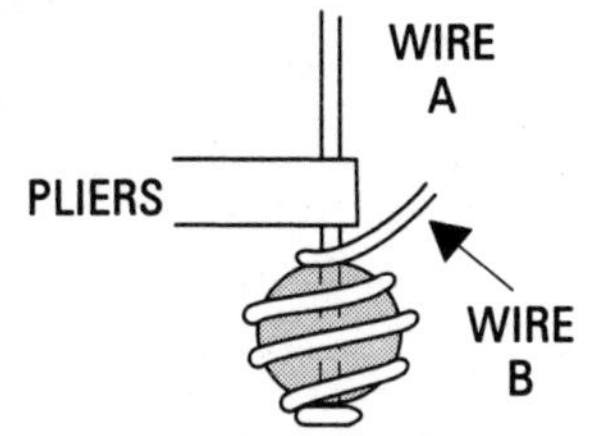

Diagram 58

G. Regardless of the size of the bead, always use a **MINIMUM** amount of wire.

H. Wrap wire B (after it is at the top of the bead) around wire A in 3 tight coils, and cut it off close. *Diagram 59*

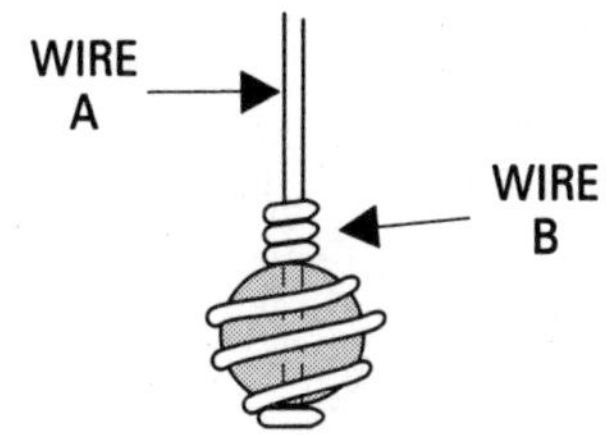

Diagram 59

I. Cut off wire A 1/2" above the wire B coils and make a loop in it. *Diagram 60*

J. Add a jump ring to this loop to give it an effective dangle, or you can add several jump rings for aesthetic value (or to accommodate taller people).

Diagram 60

SUMMARY:

You will, undoubtedly, on your first attempt at this earring, encounter some difficulty trying to spiral the wire around the bead and also wrapping those three coils at the top. If you get too frustrated, stop and take a break. Relax awhile. If you find it too difficult to use the directions given, then try to improvise with some of your own ideas. You may find a better and easier way to do things. If you use smaller beads, then use thinner wire.

REMEMBER, PRACTICE IS THE NAME OF THE GAME.

#3 EASY EARRINGS

Very little explanation is needed for these earrings. They are quick and easy to do.

A. Cut two 4" wires, 21 gauge (.028), round or square.

B. You will need four 6mm glass beads.

C. Put a bead on one end of one wire so that it protrudes out the other side enough to be gripped by the flat nose pliers. Now follow the directions on pages 33 and 34, using *Diagrams 44* through *50* for each end of these 4" wires.

D. You should now have two wires with a bead surrounded by wire on either end of each of them. See *Illustration #1* on this page.

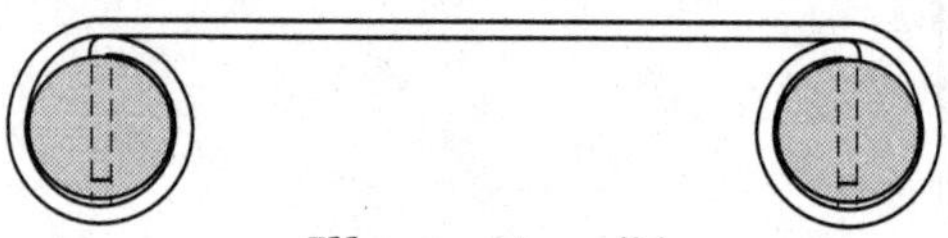

Illustration #1

E. Using the round nose pliers, grip the earrings separately and bend them as indicated by *Illustration #2* on this page.

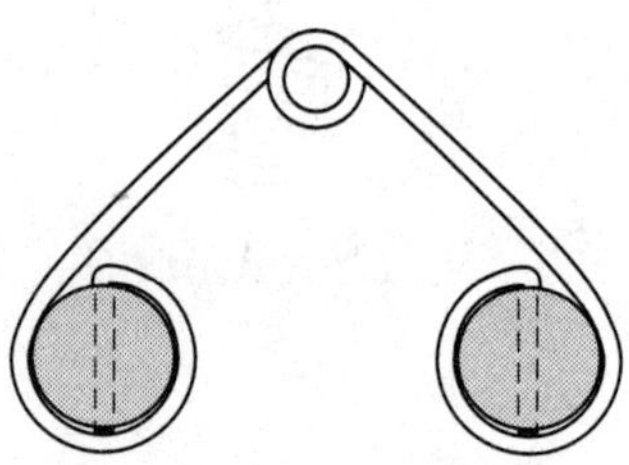

Illustration #2

F. All that you have done is make a 360 degree loop in which you will put an ear wire.

G. If you want more dangle, just make the 360 degree loop larger, and then add a small chain of three or four jump rings to it.

#1 HOW TO MAKE A COIL

To make consistently perfect coils (long or short, large or small), is very difficult to do using only round nose pliers. Why attempt this when there is a much simpler, easier and less time consuming way to do it? Some wrappers use wooden dowels which are okay, but I prefer an even easier way.

Go down to your local hardware store and check out the hardware department. More specifically, check out the machine screws and lag bolts, wood screws and sheet metal screws. They come in all sizes and lengths. The different diameters are as follows: #4, #6, #8, #10, and #12 (the larger the number, the larger the diameter). You will not need anything over 6" long. You can also use carriage bolts which come in larger diameters if you want to think really big!! I would suggest that you experiment with a few and then decide just what you need so you can carry an assortment that will serve your needs.

When you use the screws and bolts, you will be coiling from left to right wrapping snugly toward the head of the screw. If you run out of threads, simply unscrew the bolt (or screw) a few inches to make more empty threads available. To remove the wire from the bolt (or screw), simply unscrew the bolt from the coil of wire.

You now have a neat, nifty, uniform coil of wire. In the pages that follow, when instructed to make a coil, this is how it is done.

#2 FLAT COIL EARRINGS

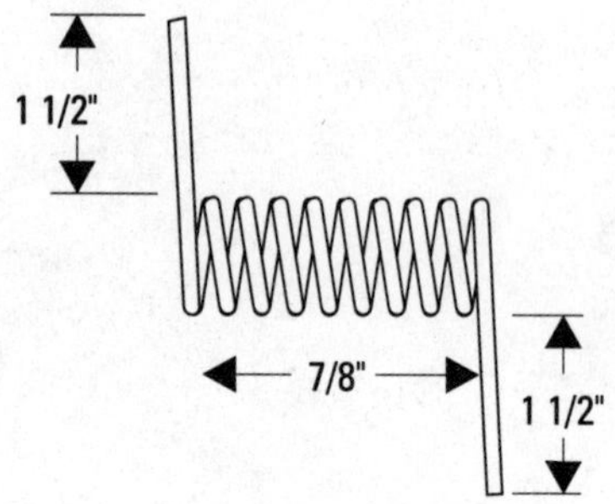

Diagram 61

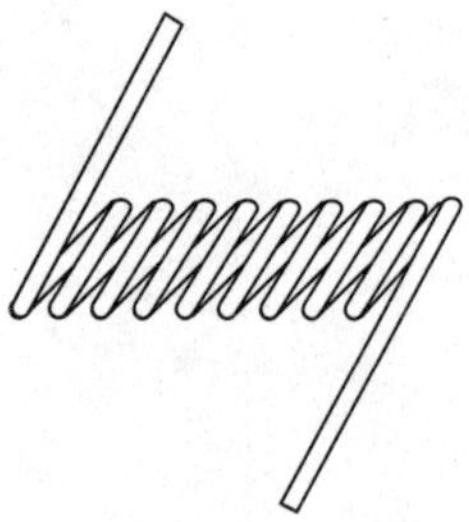

Diagram 62

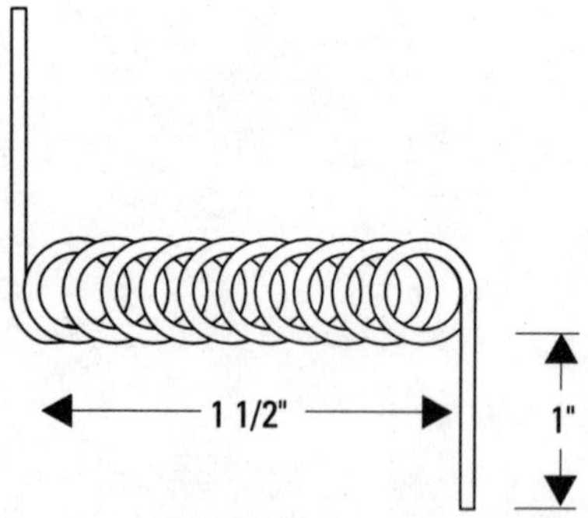

Diagram 63

A. Cut a piece of 20 gauge round brass wire 12" long.

B. Use a #8 machine bolt (or screw) 3" long for the coil.

C. Start the coil about 1 1/2" from one end of the wire, and coil it until there is only 1 1/2" of wire remaining. The ends should be pointing in opposite directions and perpendicular to the coil. Remove the bolt/screw from the coil and gently push the coils as close together as possible. The coil should be about 7/8" long. *Diagram 61*

D. Grasp the coil with the thumb and forefinger of the left hand and gently squeeze the coil, at the same time pulling your thumb to the left and pushing the forefinger to the right. This, in essence, puts a tilt to the coil. (It will take some practice to do this well.) Do not try to do the whole coil at once. Move your thumb and forefinger the full length of the coil repeating this procedure as you go.

E. When the tilt is well to the right of the vertical (90 degrees), you may then use the flat nose pliers to flatten the coils a little more. *Diagrams 62 & 63* The flattening of the coil will lengthen it. Keep this in mind when making the mate to it. They should be the same length. You may even tap them gently on the vise with the plastic hammer (turning them over after each side is tapped). *Diagram 63* If they are tapped gently with the hammer, this will make them more stable and uniform and also make them identical to each other.

F. Using the flat nose pliers, put a 90 degree bend in each of the end wires. *Diagram 64*

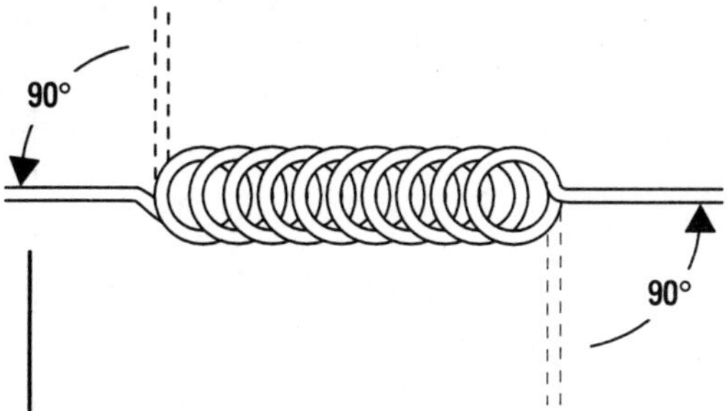

Diagram 64

G. Cut one end wire of each earring so it is 1" long. If you want only one bead on the other end, then cut it off 1" in length also. Whatever you add, be sure to leave enough wire to secure it.

H. Using the round nose pliers, finish the earring as in *Diagram 65.*

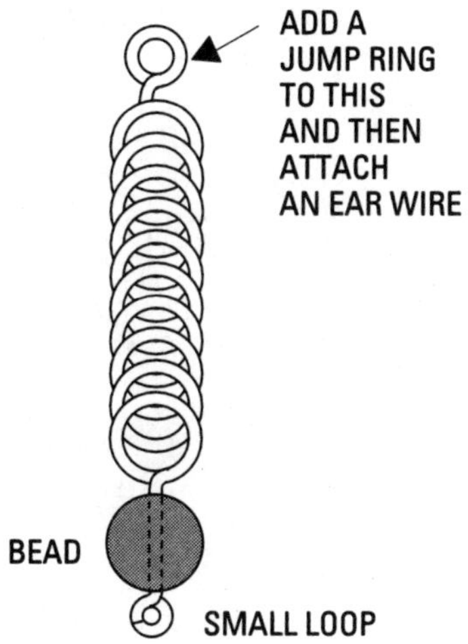

Diagram 65

SUMMARY:

After you have experimented with this system, then explore the many variations that you have (or will have) thought of. The most obvious is the simple coil earrings. You may use the directions for the flat coil except for a few minor changes. These changes are:

1. Use a #10 machine bolt to make the coil.
2. Omit steps #D and #E.
3. Stretch (very gently) the coil from 7/8" to 1 1/4".

4. Cut off one of the straight ends of each coil as close to the coil as possible.
5. Bend that end across the diameter of the coil to prevent the beads from falling out.
6. Use step #F as instructed to complete the other end.
7. Fill the coil 1/4 or 1/2 or full with 3mm or 4mm beads, making sure that the coils are close enough to each other to prevent the beads from falling out. They should not fall through the end because of the wire across the bottom.
8. Follow step #H to finish each earring. Think about what you have done, and don't be hesitant about exploring new and different possibilities. **HAVE FUN !!!**

#1 SPIRAL DISC EARRINGS

These earrings **ALWAYS** look better if they are done in twisted wire. There is no special length for them, but for the average person, they should be about 1 1/ 2" when completed. For a taller person, a 2" length looks better. These earrings are not easy to do, so it would be advisable to have some experience in wrapping before attempting these. **REPEAT**, they are not easy to do. **PATIENCE** is the main ingredient.

A. Cut two pieces of 21 gauge square brass wire about 12" long and select two 6mm drilled round glass beads.

B. Twist the wire with the pin vise.

C. Directions are for one earring. Both are identical except for the end tie-down. One of them should be tied down on the opposite side from the other so you have a left earring and a right earring. You can tell this by having the tie-down cut off on the back side of the earring. (No cut wire should be showing when the earring is being worn.)

D. Put the wire through the bead so 1/4" is sticking out the other side. Hold it firmly with the flat nose pliers and proceed to wrap the longer wire 3 1/4 times around the bead. **CONSULT DIAGRAMS 48, 49, & 50, PAGE 34, TO SEE HOW TO BEGIN AND CONTINUE THIS PROCEDURE.** *Diagram 66* shows how a completed **RIGHT** earring should look. (The cut off end of the tie-down is hidden.) When the earring is worn properly, the tie-down is toward the rear of the ear. The **LEFT** earring is tied the opposite way, but the tie-down is still toward the rear (with the end hidden).

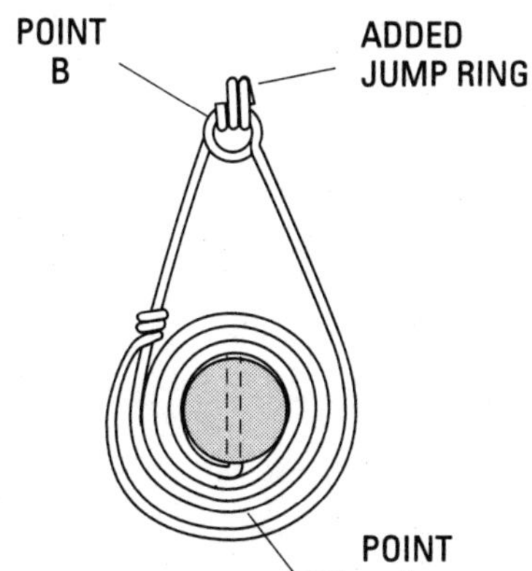

Diagram 66

The distance between point A and point B should be 1 1/2". The disc of wire around the bead, in reality, is impossible to get as tight as shown. Try to get it as tight as you can by keeping the pliers as close as possible to where you are bending. Keep the distortion of the disc to an absolute minimum.

#1 SEVEN OR NINE BEAD VICTORIAN EARRINGS

Without a doubt, beautiful things happen when you "play" or experiment with the wire and various shapes of drilled beads. Such is the case with this chapter. Once again there are many variations, and what is shown is but one of the best, most effective products of those variations. **IT IS IMPORTANT TO NOTE THAT FOR THE BEST LOOK IN THESE EARRINGS, ALL, REPEAT ALL, WIRE MUST BE TWISTED.**

A. Cut fourteen pieces of square 21 gauge (.028), brass wire, 5" long.

B. Cut two pieces of square 21 gauge or 22 gauge (.028 or .025), brass wire 4" long. These are the wrap wires, so bend over one end of these 1/2".

THIS DIAGRAM IS FOR 7 BEADS

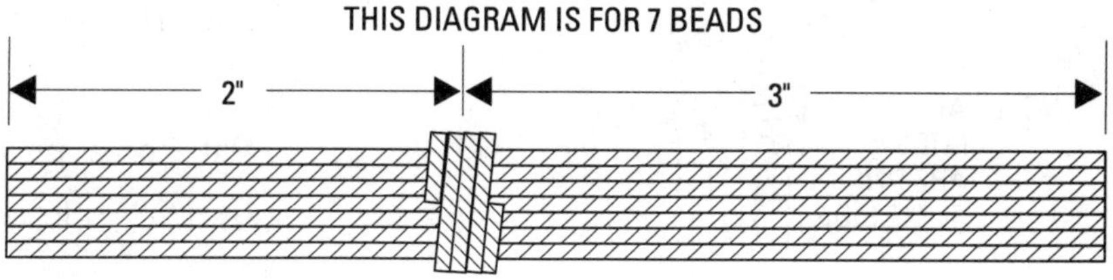

BACK VIEW
Diagram 67

C. Put seven of the 5" wires together and wrap them with one of the 4" wrap wires. Cut off excess wrap wire. Repeat this process with the other seven 5" wires and the 4" wrap wires. Now you should have two separate groups of seven wires tied together. For placement of the wrap wires, see *Diagram 67.*

D. Spread the longer wires with a wide blade pocket knife. This can be done with the fingers, but by forcing the knife between each of the wires at the wrap wire and giving it a semi-twist of the wrist, you'll get a much sharper and neater bend. You may also wonder why the wires are not twisted before you wrap them together. I can get a much tighter wrap if they are not twisted before being wrapped. It is simply **my** choice. *Diagram 67A*

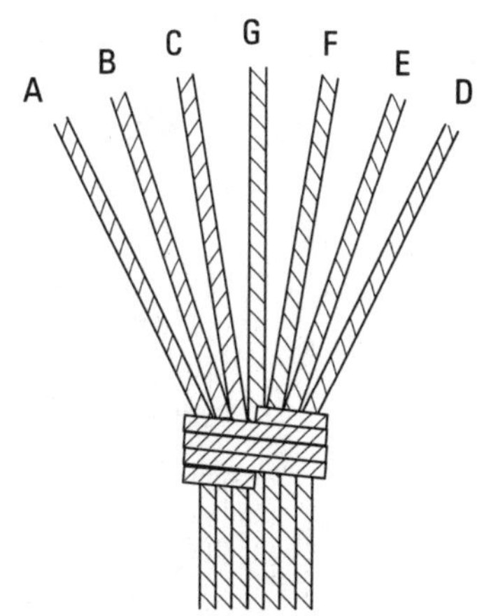

Diagram 67A

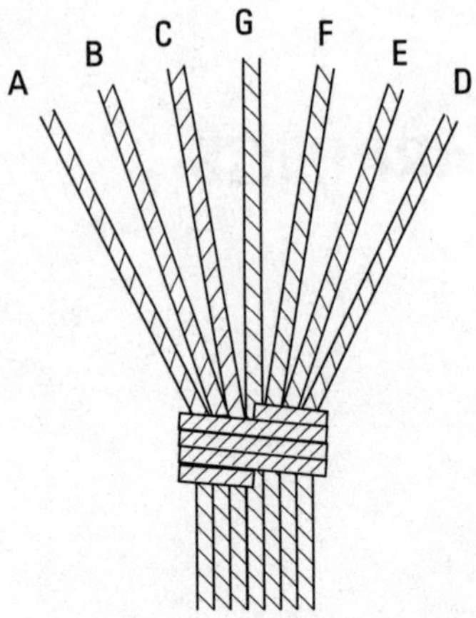

Diagram 67A

E. Select fourteen barrel-shaped beads, 6mm x 8mm.

F. Please be sure that the longer wires are spread enough to accommodate the beads. *Diagram 67A*

G. When the wires are tied together, twisted, and spread as in *Diagram 67A*, start adding the beads one at a time as indicated.

H. **NOTE: EACH WIRE IS LETTERED FOR THE FOLLOWING INSTRUCTIONS.** *Diagrams 68 & 69*

I. Put a bead on wire A so the bottom of the bead is less than 1/4" from the wrap wire.

J. Put a bead on wire B so it overlaps the bead on wire A by 1/8" or less.

K. Holding wires A and B (with the beads on them) in the right thumb and forefinger, bring wire A up the side of the bead on wire B and bend it tightly around wire B just above the bead on wire B. Cut wire A just beyond where it crosses wire B in back, and crimp it tightly to secure it to wire B. Do not crimp it tight enough to cut wire B. *Diagram 68*

L. Throughout all of this procedure, make sure **ALL CUT ENDS** are on the **SAME SIDE OF THE EARRING.** (The same side as the ends of the wrap wire.)

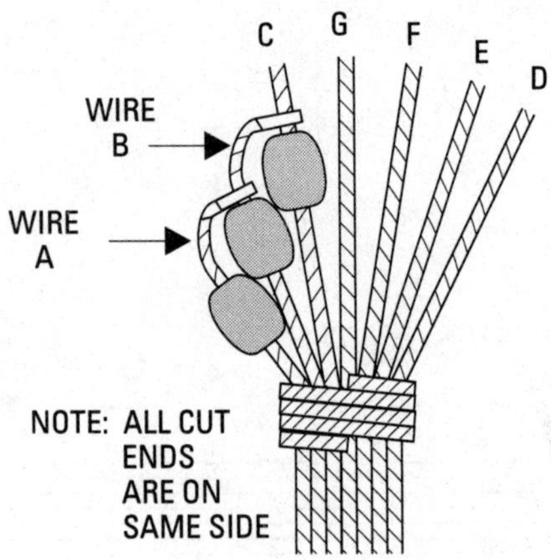

Diagram 68

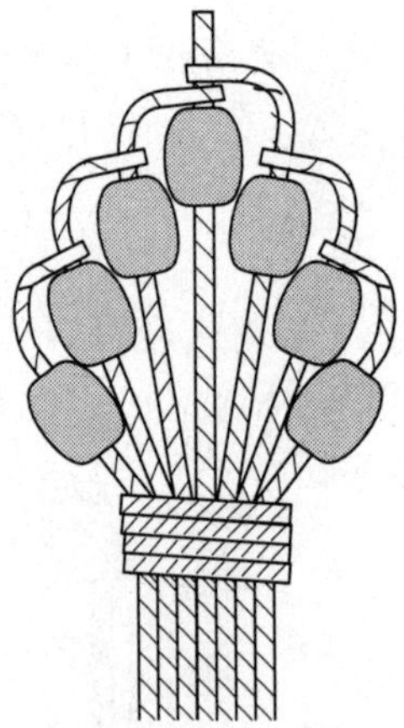

Diagram 69

M. Put a barrel bead on wire C so it overlaps the bead on wire B by 1/8" or less. Following the same procedure as in paragraph J, bring wire B alongside of the bead on wire C and across the top of the bead (on wire C), wrap it around wire C and cut it off in back after it crosses wire C. Crimp it tightly around wire C to secure it. *Diagram 68* Make sure the cut end is on the same side as the other cut ends.

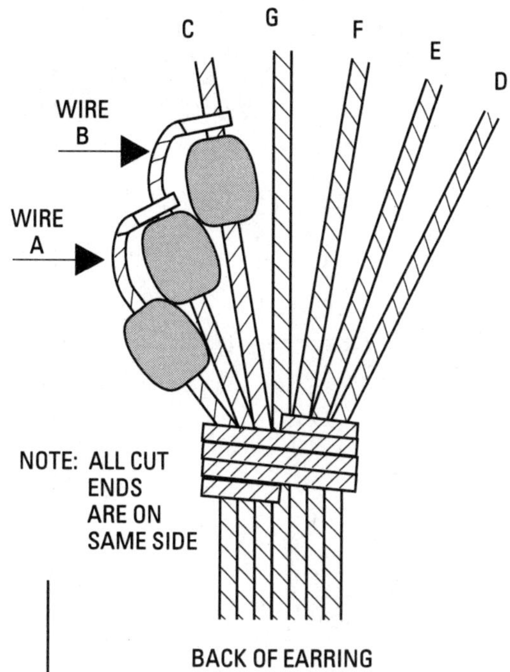

Diagram 68

N. Now do likewise with wires D, E, and F making sure that all cut ends are on the same side as the cut ends of the wrap wire.

O. Put a barrel bead on wire G so it overlaps the bead on wires C and F by 1/8" or less.

P. Bring wire C up alongside of the bead on wire G and across wire G and immediately wrap it around wire G and cut it off just beyond wire G. Crimp it tightly around wire G.

Q. Bring wire F up alongside of the bead on wire G and across the top of the bead on wire G and above wire C (which is already fastened). Wrap it around wire G and cut it off (as you did with wire C). Crimp it tightly around wire G. *Diagram 69*

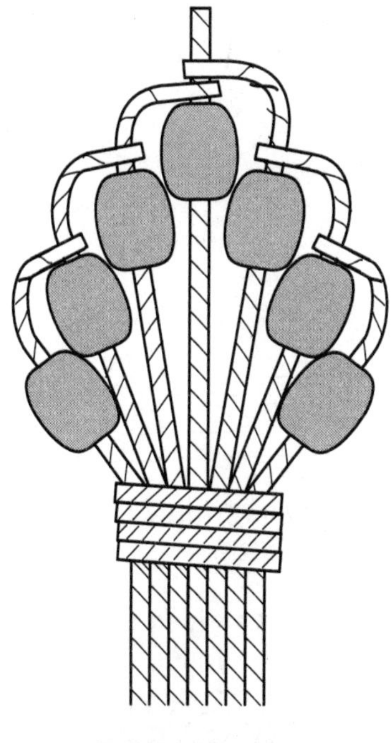

Diagram 69

R. It is optional as to what can be done with wire G. It can be cut so it protrudes 3/16" past the crimped C and F wires, then made and flattened so it faces the back, or longer length can be made and swirled into a decorative disc, or you can put a contrasting color bead on it and secure it.

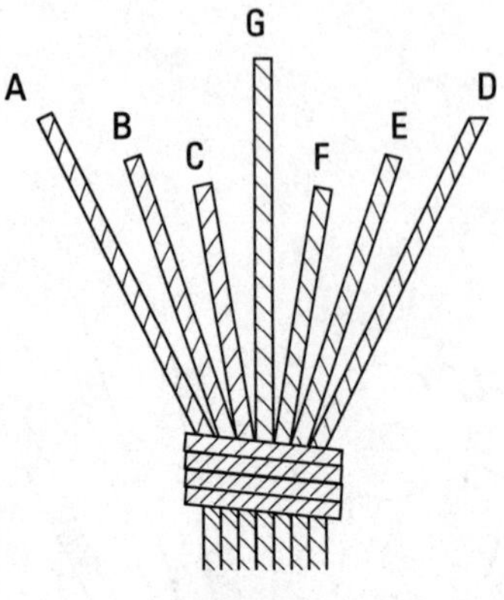

Diagram 70

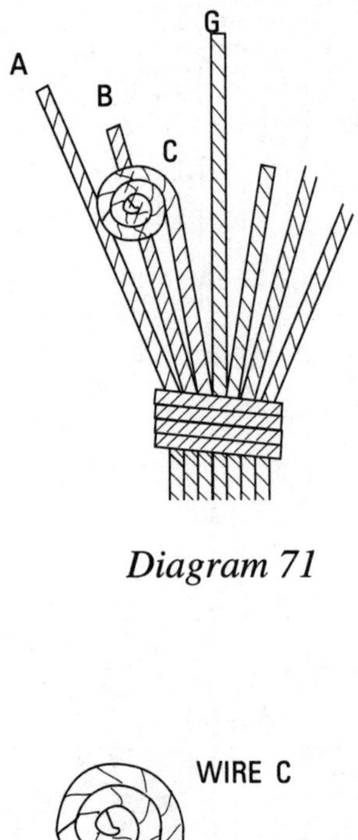

Diagram 71

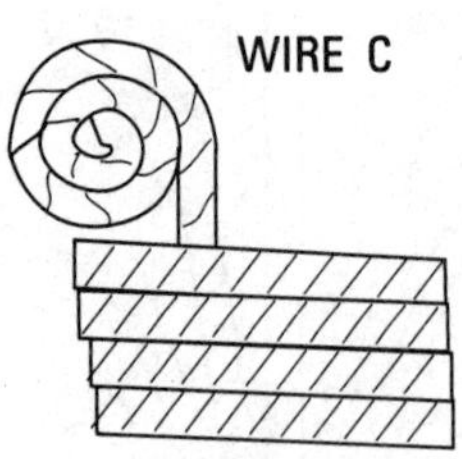

Diagram 72

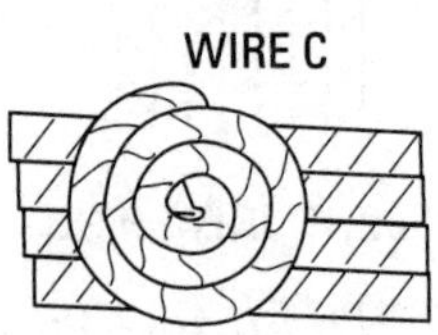

Diagram 73

S. There are many ways to finish this earring. What follows is a result of my own experimentation. It looks great when completed.

T. Spread the wires as indicated in *Diagram 70.* **DO NOT CUT WIRES A, G, or D.** Cut wires B and E 1/4" shorter than wires A and D. Cut wires C and F 1/4" shorter than B and E. The lengths are arbitrary. Experiment to suit your own needs.
JUST REMEMBER, WIRES A, G, and D SHOULD BE LEFT LONGER THAN THE REST OF THEM.

U. Beginning with wire C and using the smallest end of the tapered round nose pliers, curl the end of the wire counterclockwise to form a small circle. (The front of earring must be facing you. The cut ends are in back). Using the flat nose pliers, continue making the circle so it becomes a flat disc. Use up all of the wire and pull the disc down so it covers part of the center portion of the wrap wire. *Diagrams 71, 72, & 73*

V. Do the exact same thing with wire F, only put it on the opposite side from wire C in the exact same position on the wrap wire.

W. Do the exact same thing with wire B. Be sure to put it on the same side as wire C. Do not put it directly over wire C when you finish. Put it so it covers more of the wrap wire to the left of wire C.

X. Do the exact same thing with wire E, and when it is finished, put it on the same side as wire F, covering the wrap wire to the left of wire F.

Y. Do the exact same thing with wire A, and put it on the same side as wires B and C so it covers the remainder of the wrap wire.

Z. Do the exact same thing with wire D and put it on the same side as wires E and F and cover the remainder of the wrap wire.

AA. Using the large part of the round nose pliers, roll wire G into a large (or small if you prefer) jump ring to hang on an ear wire.

See color photo on cover or insert for finished piece.

SUMMARY:

After you have made a few pair of these multi-bead earrings, you will have discovered that the sky is the limit for this design. I have put as many as sixteen and as few as three small oval beads in one pendant (See cover or insert). The variations of this design are only limited by your imagination. You may even want to mix brass and copper wire together. It looks great.

Small
Birdcage Earrings
Triple Stone
Earrings
Pendant and
Earring Set
Small
Victorian Earrings
Earring
Variation
Multi-Bead
Pendant
Elevated
Oval Cut Ring
Victorian
Earring Variation
Flattened
Coil Earrings
Open Birdcage
Double Bead
Earrings
Various Sizes of Birdcages

Copper Coil Bracelet, Matching Ring and Pendant
Triple Mother of Pearl Pendant and Earrings
Birdcage Tandem Earrings
Coil Earrings
Simple Spiral Earring
Victorian Earring Variation
Birdcage Pendant
Triple Birdcage
Victorian Earrings
Brass and Glass Bracelet

#1 OTHER FANCY EARRINGS

These earrings are simple and easy to make. The finished product can be seen on the color insert. There is no limit to variations. A minimum of three wires is required and the most effective maximum is nine wires. For symmetry and balance, **ALWAYS** use an odd number of wires.

The center wire is always used for the hanger wire or the pin (for a brooch). A very radical approach would be to use thirteen, fifteen, or seventeen wires without beads and arrange the swirls not one after the other, but in a staggered circular arrangement. The effect is absolutely stunning. Also keep in mind wire lengths. I have found that a good minimum length is 3" (for earrings), and a good maximum length would be 10" (for a pendant). What follows in this chapter is the simplest form of this design. Have fun and enjoy!!!

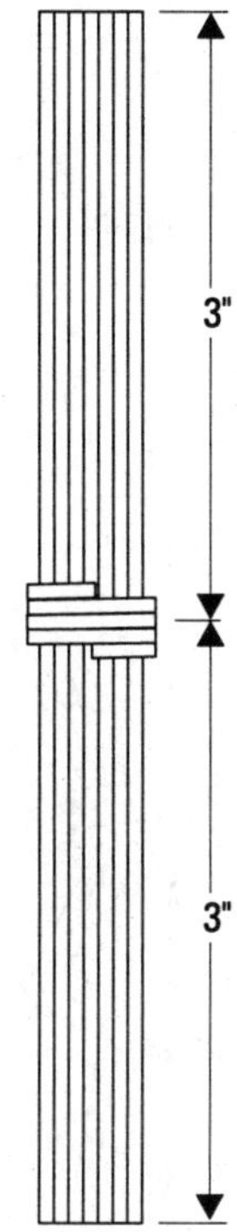

Diagram 74

A. Cut seven wires (brass, copper or mixed) 21 gauge (.028) square, 6" long.

B. Cut another wire 21 gauge (.028), square, 4" long and bend over one end 1/2".

C. Put the 6" wires together, and wrap them with the 4" wrap wire as shown in *Diagram 74.* Cut off excess wrap wire.

D. For a more pleasing design, twist one 3" section of wires one at a time. Leave the other section plain. *Diagram 75*

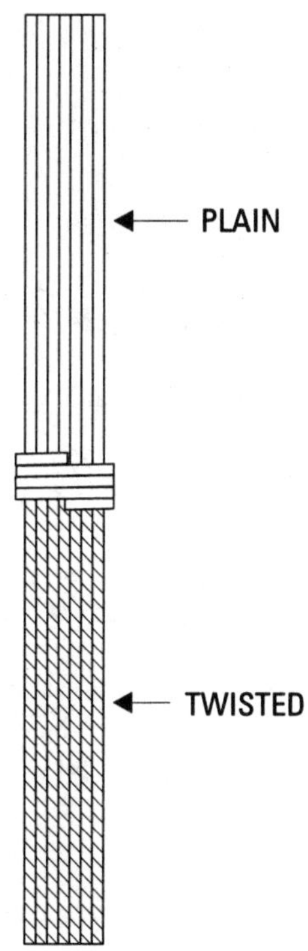

Diagram 75

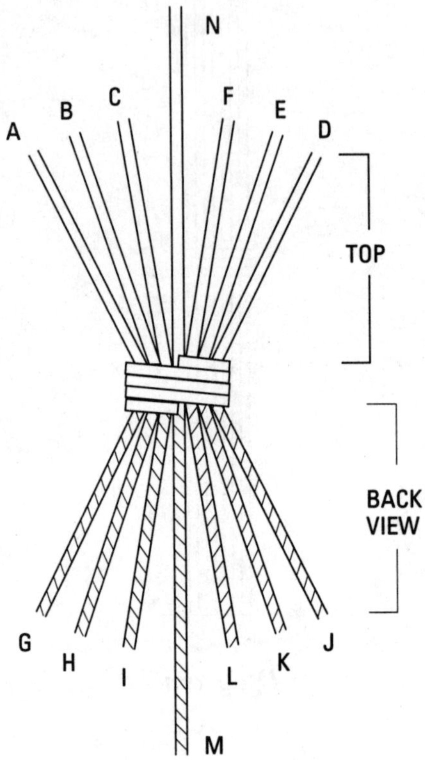

Diagram 76

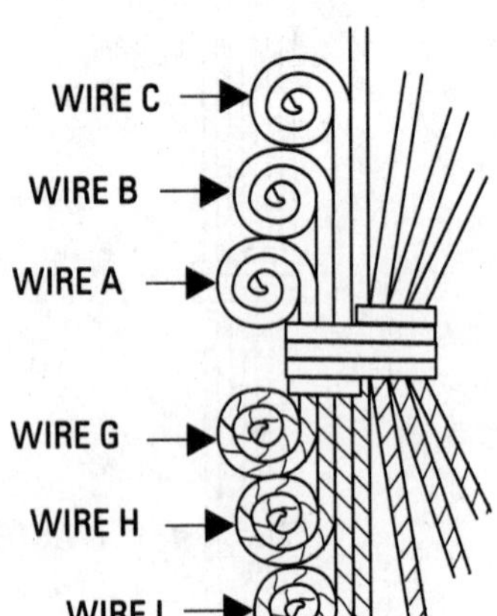

Diagram 77

E. Even though *Diagram 75* shows the wires very tight, in reality, these wires are spread a little bit. (This happens because the pin vise, used in twisting, is forced on each wire.) This spreading can be held to a minimum if each wire, when twisted, is bent up **SLIGHTLY** so the pin vise will not touch the wires on either side of the wire being twisted. Bend each wire back down separately when it is twisted.

F. Using the wide blade of a pocket (or kitchen) knife, spread all of the wires slightly. Once again each wire is lettered for instructional purposes. *Diagram 76*

G. You do not have to spread the wires this much. It is drawn this way for clarity and labeling.
NOTE LETTERING OF THE WIRES.

H. Proceed as follows: Cut 1" from wires A, G, D, and J so that they are 1" shorter than the other wires.

I. Using the small end of the round nose pliers, curl the end of the wire away from the vertical center (of the group of wires) to form a small circle. This is for wires A, D, G, and J only. Then with the flat nose pliers holding the small circle of wire in the edge of the jaws, push the rest of the wire around it. Rotate in small bites until all the wire is used and the now created flat disc is touching the wrap wire. *Diagram 77*

J. Then do likewise with wires B, H, C and I as indicated in *Diagram 77.* **DO NOT CURL ANY OF THE OTHER WIRES DOWN TO THE WRAP WIRE. CURL THEM ONLY TO THE PRECEDING WIRE**. *Diagram 77*

K. Now repeat paragraphs H, I, and J with wires D, J, E, K, F and L.

L. Using the pin vise, twist wire N. This will prevent it from being accidentally pulled out.

M. Put a bead (round or long) on wire M and slide it up so that it touches wires I and L. At the bottom of the bead make a small circle of wire (with the round nose pliers) to prevent the bead from slipping back down.

N. With the round nose pliers, make a circle of wire with wire N. Keep in mind, this is to be used to attach an ear wire.

O. A totally different effect can be obtained by putting the wrap wires closer to one end and then making the discs in a counterclockwise swirl (instead of clockwise).
See the red earrings in color photo on cover or insert.

#1 COMPLEX ALL WIRE RING

It is amazing what can be done with wire only jewelry, particularly with a mixture of brass and copper wire (or even with gold and silver wire). By arranging different wires in different orders, beautiful symmetrical patterns of color emerge to enhance the original design. The only limit is the imagination.

The ring that is being dealt with in this chapter is a perfect match for the all wire pendant and earrings. It is the **SAME IDENTICAL SYSTEM,** except the ring requires shank wires to go around the finger. This is the only difference except for the amount of wire used in each, and that is a matter of taste. In this chapter there will be seven wires used in the ring, five wires used in each of the earrings, and thirteen wires used in the pendant. For the complex all wire ring here is what you will need:

Materials: Four copper wires (.028, square) 5" long.
Three brass wires (.028, square) 5" long.
Two brass wires (.028, square) 3 1/2" long.
Bend these over 1/2" on one end 180 degrees.
One brass wire (.028, square) 4" long.
Bend over one end 1" 180 degrees.

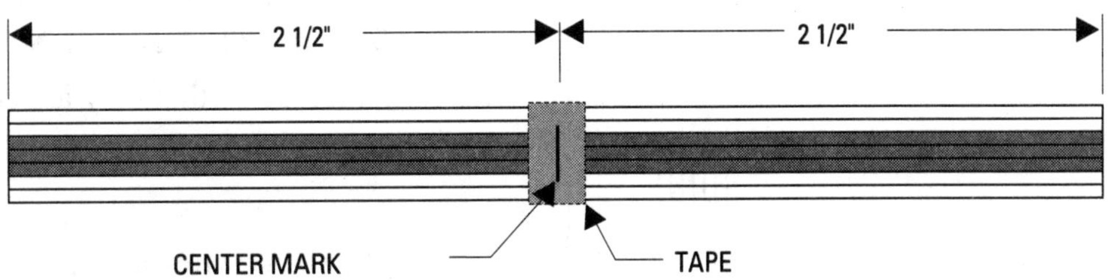

Diagram 78

A. Put the 5" wires together (you may want to put the three brass wires in the middle, but that is optional). Tape together, snugly, parallel to each other (with paper tape for less residue), and mark the center of the wires with a felt tip pen. *Diagram 78*

B. For a size 6 ring, mark 3/4" on each side of the center mark for placement of the wrap wires. For every two sizes larger, move the wrap wires out 1/16" more on either side. (Example, size 8: 13/16 of an inch, size 10: 14/16 of an inch and so on.) For a size 6, put the wrap wires as indicated in *Diagram 79*.

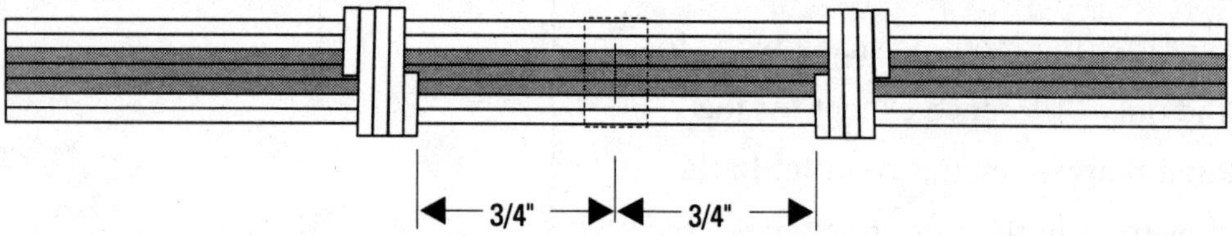

Diagram 79

C. For a size 6 ring, bend the wires around the size 5 part of the ring mandrel. *Diagram 80* Please note that after the wires have been bent around the mandrel, the wrap wires **MUST** be directly opposite each other, and the **CUT ENDS SHOULD BE ON THE INSIDE.** *Diagram 81*

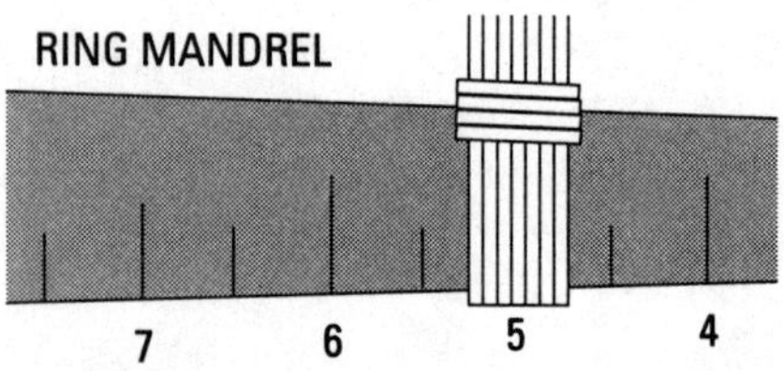

Diagram 80

D. Remove the entire assembly from the mandrel and using a wide blade, turn up the outermost wire on each side. *Diagram 82* Push, **GENTLY**, the wires you have just turned up so that they are completely vertical and form a circle with the shank wires.

E. The shank wires are the wires that go around the fingers. Because of the two that were turned up, there should now be five shank wires. *Diagram 83* The next step is extremely important so proceed slowly. How well it is done determines the accuracy of size and the degree of roundness of the ring.

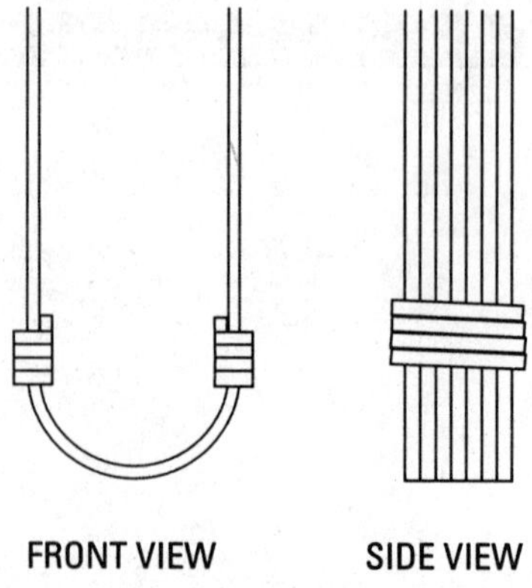

Diagram 81

F. Holding the ring with both hands (between the forefinger and the second finger) so that the thumb of each hand is directly over the wrap wire on either side, place the bottom of the shank wires on a solid flat surface, vertical to it, and push down **GENTLY**, with the thumbs or thumb nails on the wrap wires. Push the entire assembly down so the ring is much smaller than a size 5. Now put the ring back on the mandrel and force it to a size 5. If done properly, the ring should now be a perfect circle (size 5). Do not worry about the small size. When this step is completed, it can be stretched. Now spread the wires that come out of the wrap wires on either side. *Diagram 84*

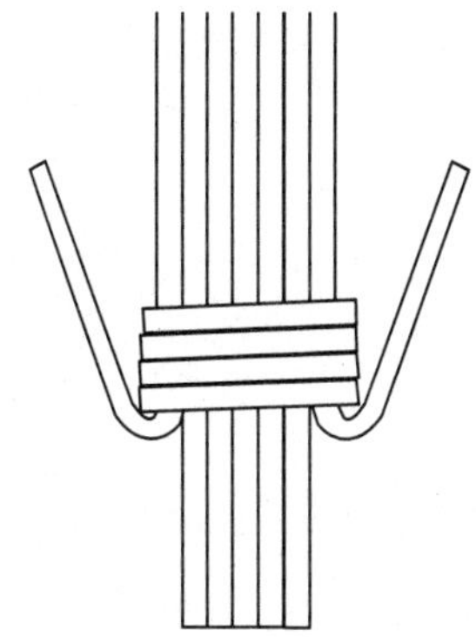

Diagram 82

FRONT VIEW AFTER TURNING UP OF OUTSIDE WIRES

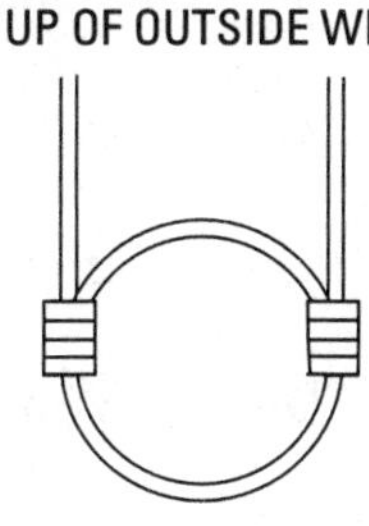

Diagram 83

G. The wires that were turned up earlier (in paragraph D) must be spread slightly to accommodate what is to come. *Diagram 84* shows only one side of the ring--remember, there are now 14 wires to deal with (7 on each side). These wires must be placed the same as if you were putting your hands together, fingers first, and each finger alternating left, right, left, right, etc.. For instructional purposes, the wires will be numbered 1 through 14. Put the ring back on the mandrel (on size 5). The wires on the left will be the odd numbers, and the wires on the right will be even numbers. Interweave them in **NUMERICAL** order:

Number 1 wire from left to right across the top of the mandrel and push the end down slightly. While doing this you **MUST** hold and press the assembly at the wrap wires with the left thumb and forefinger right at the size 5 area of the mandrel.

Number 2 wire from right to left, etc.

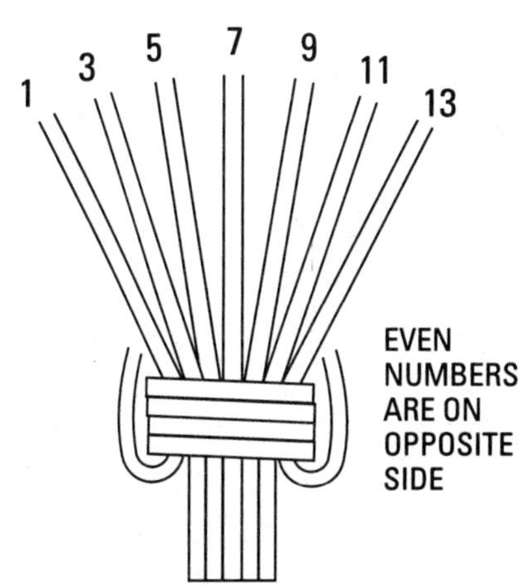

Diagram 84

Number 3 wire from left to right, etc. and so on until all 14 wires are in their proper place. (As you have discovered by now this procedure will really decide whether you are sincere about becoming a serious wirewrapper.)

THIS SYSTEM IS DIFFICULT BUT IT CAN BE DONE WITH PATIENCE, PERSEVERANCE AND PRACTICE.

You may not do well the first time or the second time, but persistence really pays off the third or fourth time when the ring really begins to come into its own when paragraph G is completed.

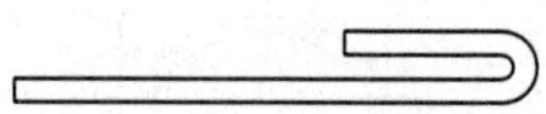

Diagram 85

H. You should have 14 wires across the top of the ring between the two wires that were turned up with the knife blade in paragraph D. The 4" wire cut and bent earlier can now be used. *Diagram 85*

I. Hook the entire top of the ring with this wire making sure that the small end of the hook is across the top and push the long end up so it is vertical to the assembly. Squeeze the hook, gently, with the flat nose pliers, and while it is still in the jaws of the flat nose pliers, insert the small tip of the hook into the vise and tighten it (the vise). Then grip the longer wire with the flat nose pliers, pull it toward you and down firmly. This, in effect, completes one wrap around all of the wires in the center of the ring. Make sure the interwoven wires are all neat and orderly between the wires that were turned up. Wrap this wire around two more times (making it very snug), then cut off the excess of the wrap wire.

J. The top of the ring should look like *Diagram 86* . It is enlarged for clarity, and the wires are numbered.

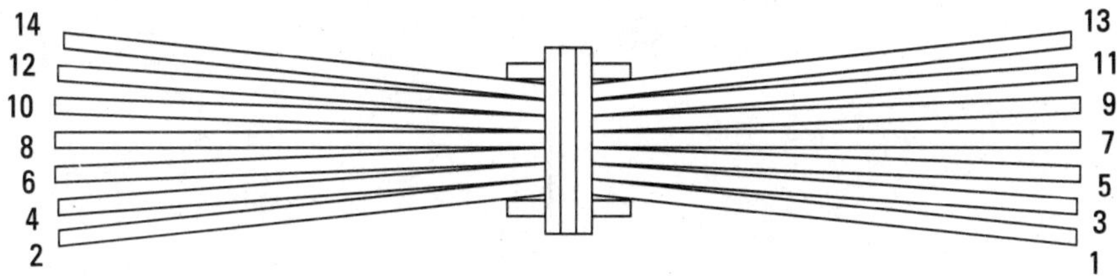

Diagram 86

K. Check the ring for size. If it is too large, then begin your next attempt 1 1/2 sizes smaller than the desired size. If it is too small, then it can be stretched with very little adverse effect. **PRACTICE!!!**

L. At this point there are many options--just keep in mind that all wires will be rolled into flat discs, and all 14 of them will occupy a 1/2" circle or a 3/4" oval right on top of and around the exposed wrap wire. *Diagram 87*

TOP VIEW OF RING
Diagram 87

OPTION #1 Cut 1" off and twist every other wire, and roll them all into a small circle in the middle (on top of the wrap wire), then roll the rest into flat discs and surround the ones in the middle.

OPTION #2 If brass and copper wire has been used, put brass in the middle and surround them with copper coils.

OPTION #3 Cut all of them various lengths and arrange the coils at random.

M. You may want to make the 3 center wires 1/2" shorter so they can be rolled and put in the center.

N. Regardless of choice, when the ring is completed, put it on the mandrel and tap it gently so all coils lie flat and snug, touching each other.

See color photo on cover or insert for finished piece.

#1 ALL WIRE PENDANT AND EARRINGS

A. These directions are for the pendant **AND** the earrings. The only difference is the length and number of wires being dealt with.

B. It is also important to note that when making earrings, in order to have them reasonably well matched, they must be made together; I would advise this set be made together so that each wire, when coiled, can be put in the identical same place on the other earring. Just alternate each earring with the other, wire by wire for accuracy of design and matching similarity.

C. There will be no separate directions for the earrings because they are made exactly like the pendant. There are, however, other options for the pendant. For a small pendant (approximately 1" in diameter when finished) use thirteen wires 6" long. For a larger pendant (approximately 1 1/2" in diameter when finished) use thirteen wires 10" long. If a larger pendant is desired, increase the length of the wires by 5". Experiment! It's a lot of fun!!! *Diagram 88A*

13 OR 15 WIRES
TIED TOGETHER
WITH A WRAP WIRE

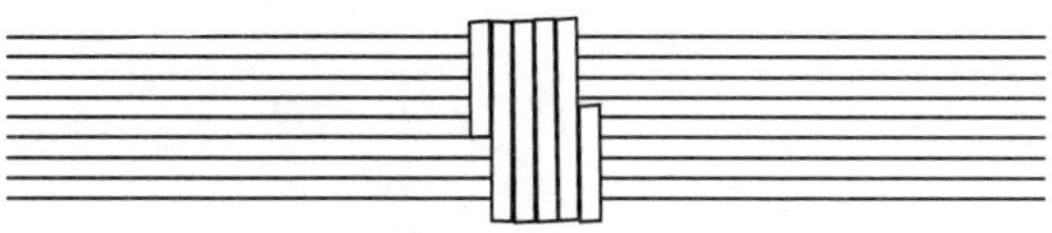

Diagram 88A

D. For design appeal, different textures intermixed are very desirable; therefore random twisting of some of the wires (before being coiled into a disc) is a must. It is a good contrast when they are interspersed with plain coils, especially when arranged in a circle.

E. It is also important to note that there is no front or back to this pendant.

F. Another appealing design would be to put 2mm beads at the center of a few or all of the flat disc coils.

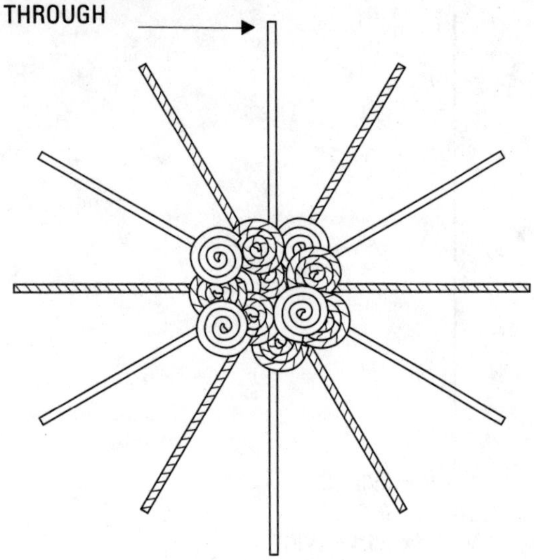

Diagram 88B

Diagram 89

See color photo on cover or insert for finished piece

G. It is also important to note that the wires closest to the center of the group can be shortened 1/2" to 1" before being twisted and coiled into flat discs.

H. As the coiling process begins, start with the inside wires and cover up the wrap wire (on **BOTH** sides) first. This sets the pattern for the entire pendant. In the center of the pendant there will be 6 flat discs on one side and 7 on the other side spaced so as to cover up the wrap wire. (There will be 7 on one side because the exact center wire is the other end of the wire to be utilized as a hanger which will be made into a round jump ring for a chain to pass through).

I. Note the photo of this pendant on the cover or insert to get a better idea of what you are to do. When the center wires are completed, you now have a half-finished pendant. It looks like a small round "Oreo cookie", part of it consisting of 6 flat discs on either side with the middle "goodies" consisting of the wrap wire from which radiate the rest of the wires. *Diagram 88B*

J. The reverse side looks identical except for 7 discs as explained earlier. The remaining wires are shown as every other one twisted. These wires can be rolled as indicated and placed tightly between the sandwiched coils already completed. *Diagram 89*

K. Just a reminder: Earrings are done the same way, just on a smaller scale.

#1 EMERALD CUT MOUNTING

This wrap is designed specifically for a large rectangular emerald cut stone. This is not an original design. One of these wraps was brought to me to be redone (after years of wear). This design is gaining in popularity so that is why it is being included in this book. The difficulty of this procedure is minimal, and it does look somewhat impressive when completed. I suspect that this design evolved as an alternative to the more difficult Tiffany mount system. Even though this book deals with brass and glass, this system is included because of popular demand. I have only done this using large rectangular smoky quartz and 20 gauge (.032) gold filled wire. It can be done using any kind of wire and any emerald cut stone. Have fun and go for it!!!

FOR A BRASS MOUNT, SIZE 6:

A. Cut four pieces of wire, (.028, square brass) 5 1/2" long.

B. Cut two pieces of wire, (.028 square brass) 4 1/2" long and bend over one end of each 180 degrees 1/2".

C. Put the longer wires together in a group, tape them together in the middle, and mark the exact center. *Diagram 90*

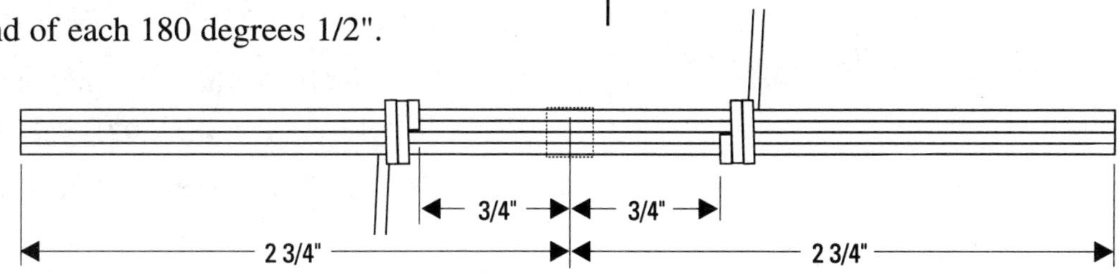

Diagram 90

D. Measure 3/4" from center mark and wrap as shown. (For larger than size 6, move wrap wires out an additional 1/16" on each side for every two sizes larger.)
DO NOT, DO NOT cut off excess wrap wire.
Remove tape from group wires.
PLEASE NOTE THAT THE WRAP WIRES ARE IN THE OPPOSITE DIRECTION. THIS IS ABSOLUTELY NECESSARY FOR THE NEXT STEP.

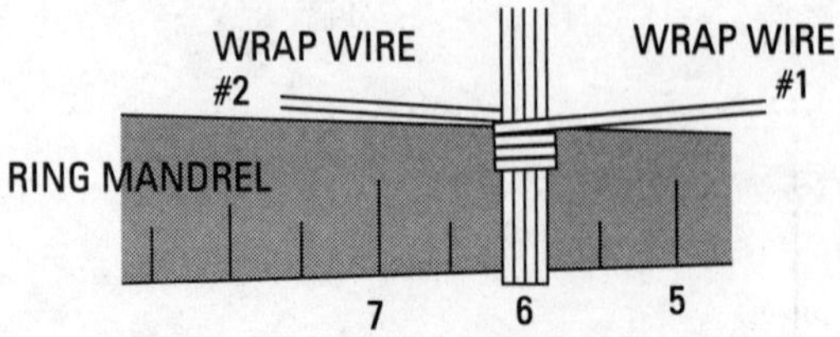

Diagram 91

E. Using the ring mandrel, bend the group wires (shank wires) around that portion of the mandrel that indicates size 6. *Diagram 91* Also make sure the wrap wires are on the outside, and the cut ends are on the inside.

F. With the wires still on the mandrel, bend wrap wire #1 across the top of the mandrel and bend it around the wires coming up the other side of the mandrel just above wrap wire #2. Wrap it neatly and flat against the shank wires one complete turn. Do likewise with wrap wire #2 only in the opposite direction. If this procedure is done properly, you now have a size 6 ring.

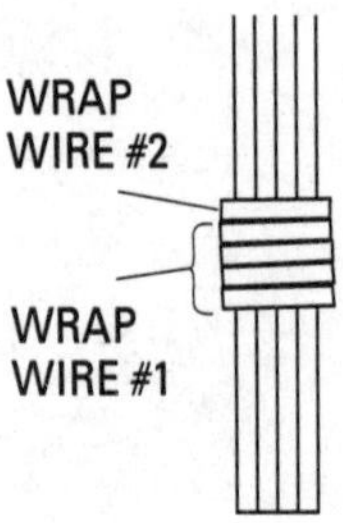

Diagram 92A

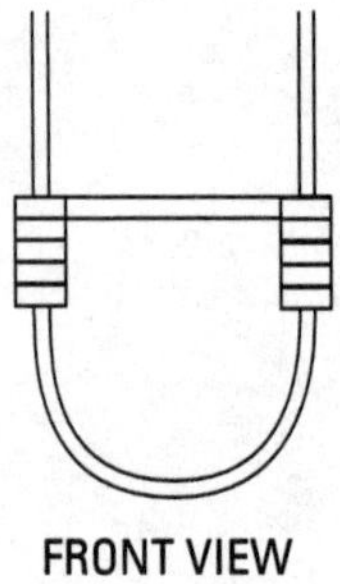

Diagram 92B

DO NOT CUT OFF EXCESS WRAP WIRE YET.
Diagrams 92A, 92B & 92C

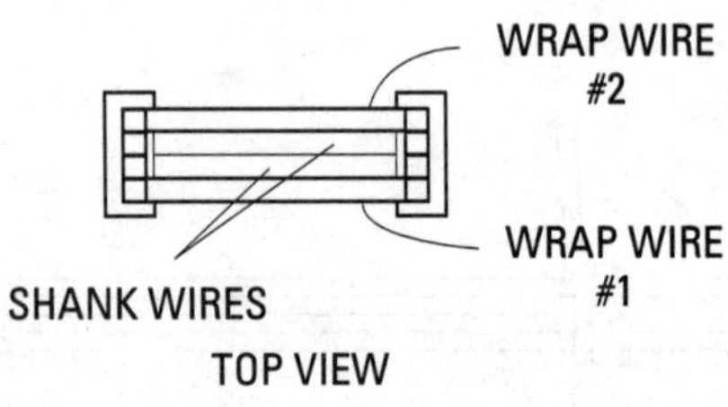

Diagram 92C

G. Measure the stone to see how high the pavilion is (how thick it is). *Diagram 93* This distance determines how high the wrap wires must be wrapped to accommodate the stone. The wrap wire must be as high as the facet cut on the pavilion. *Diagram 94*

Diagram 93

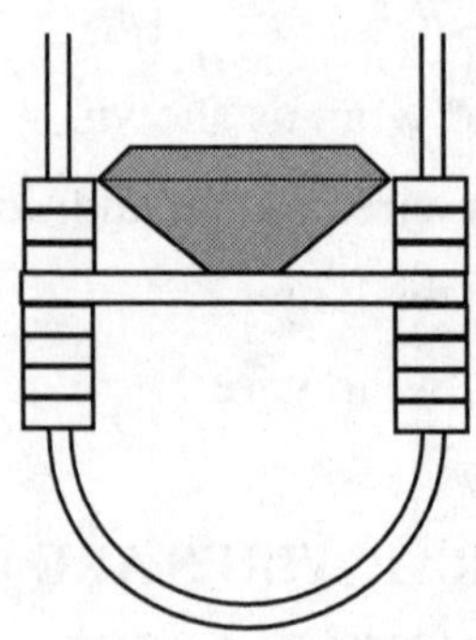

Diagram 94

H. Spread the shank wires in groups of two where they come out of the wrap wires. *Diagram 95*

I. To mount the stone proceed as follows:

Hold the stone on the two wires wrapped around the shank wires--making sure it is centered on each side and front and back. *Diagram 96*

Each group of 2 tie-down wires will be treated as **ONE WIRE**. Holding the stone as indicated in *Diagram 96*, begin with corner A as seen from the top view *Diagram 97* -- corners are now lettered. Bend the two wires across the corner of the stone and then down and to the front center of the stone. Continue downward and across to one of the wires on which the stone is resting. Tie down on the opposite of its origin. Cut the two wires you have just bent so you have just enough left to go around that anchor wire. Crimp it tightly for the tie-down. *Diagrams 97 & 98*

Do likewise with corner C next, and follow the same procedure.

Do likewise with corner B.

Do likewise with corner D.

Check *Diagrams 97 & 98* for accuracy of directions.

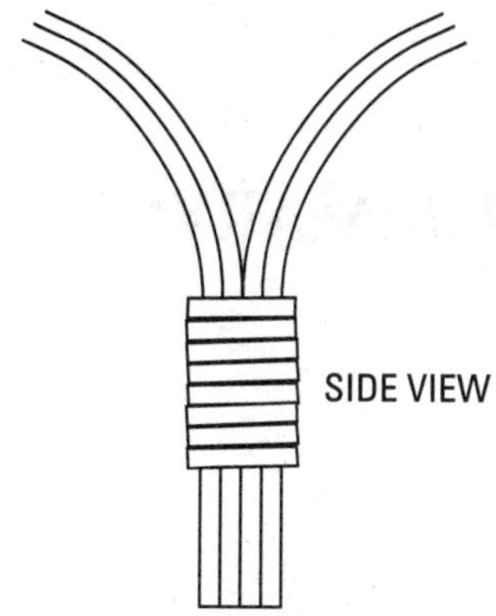

Diagram 95

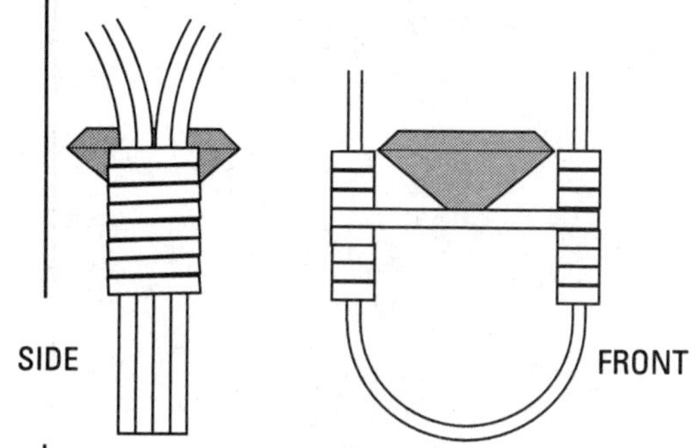

Diagram 96

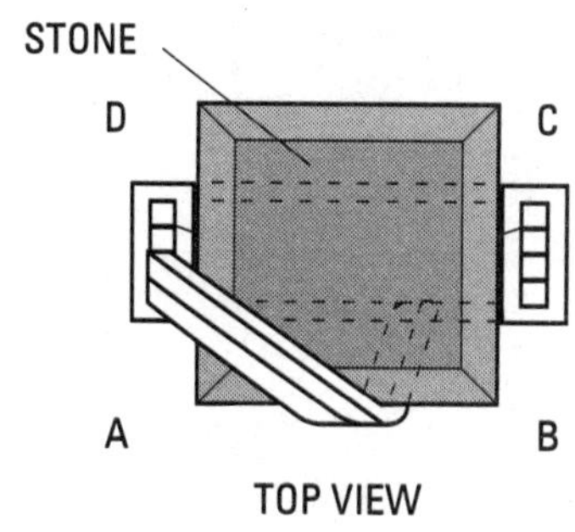

Diagram 97

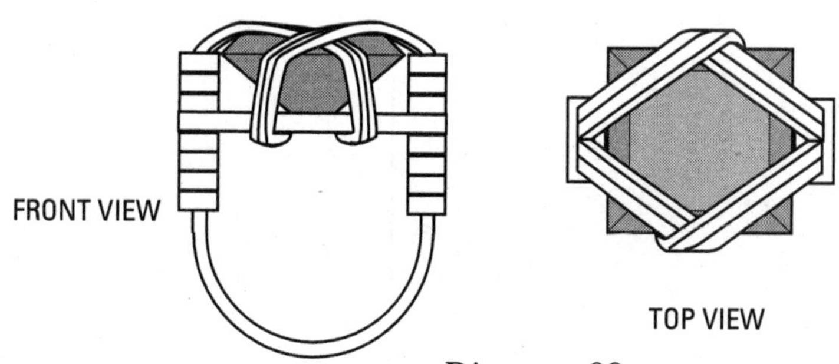

Diagram 98

CHAPTER 14 RING, EMERALD CUT MOUNTING

SUMMARY:

This is a fun ring to make after you get used to the system. Just proceed slowly, and do a neat job. If you are into serious jewelry making, put a couple of these in your display case. They will draw a lot of attention. This design can also be used on large baroque stones. Even though the symmetry may not be the same, you will discover more of your own creativity by ad-libbing for balance of design (even using beads).

#1 ELEVATED OVAL CUT STANDARD FORM RING

This wrap brings out the best in a faceted oval cut high pavilion 18 x 13mm stone. It is not an original design. Every wirewrapper has a different method, but this way has been well received by the public. HAVE FUN !!!

FOR A BRASS MOUNT, SIZE 7 1/2:

A. Cut twelve wires, (.028, square brass) 5 1/2" long. Cut two wrap wires 4" (.028 square brass) and bend over 180 degrees 1/2", of one end of each.

B. These directions are for a size 7 1/2 ring. Because of the high pavilion of the stone, it is going to be raised up high enough so the point on the bottom of the stone will be well above the finger so as not to irritate it. If you are familiar with the Men's standard form ring, (and you should be at this point), you know that to accommodate the stone, two outside shank wires, one on either side, are turned up to cradle the stone between them. The same is done here; however, to raise the stone above the finger, a third wire is turned up at a particular point in the process. When this wire is turned up (on both sides), it automatically shrinks the ring 1 1/2 sizes and at the same time raises the stone well above the finger. Keeping this in mind, proceed as follows:

C. Put the twelve 5 1/2" wires together, parallel to each other, taped at the center. Mark the exact center (2 3/4" from either end). These are the directions for (at this point) a size 9 ring.

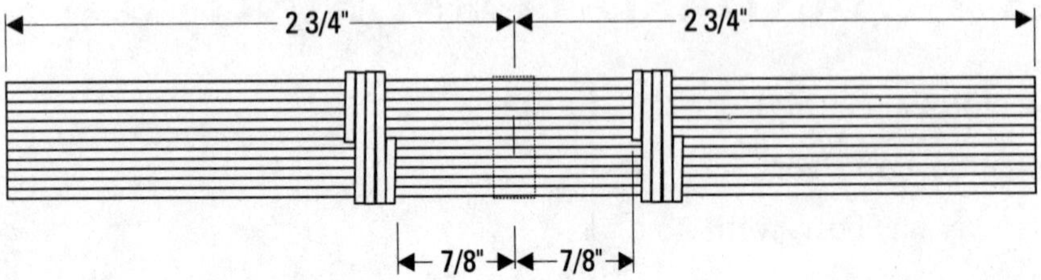

Diagram 99

D. From the center mark, measure out from either side 7/8". This is where you will put the wrap wires on either side. *Diagram 99*

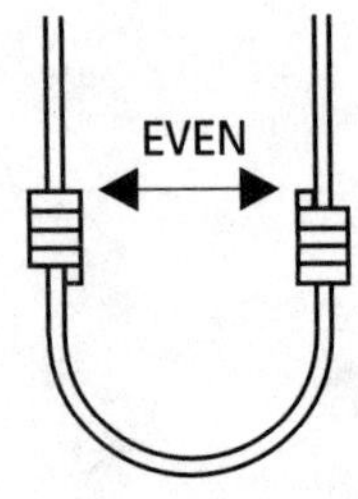

Diagram 100

E. When the wrap wires are completed, remove the tape and bend the wires around the size 7 portion of the mandrel. **MAKE SURE THE CUT ENDS ARE ON THE INSIDE**. It is also important to make sure that when the bend is complete, the wrap wires are directly across from each other. *Diagram 100*

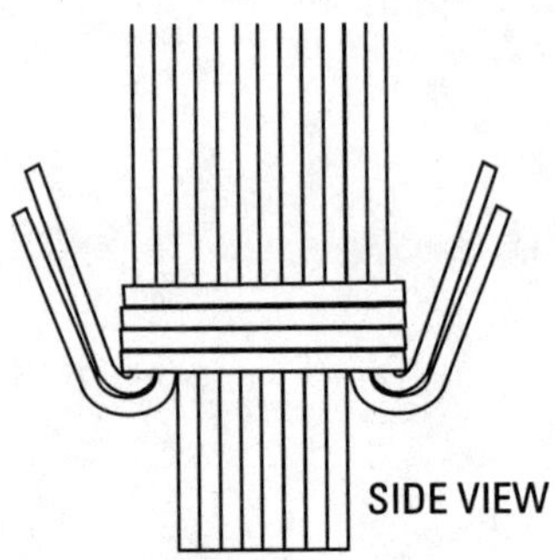

Diagram 101

F. With a wide blade knife, turn up the two outside wires on either side. *Diagram 101* Also make sure the assembly is right on size 9.

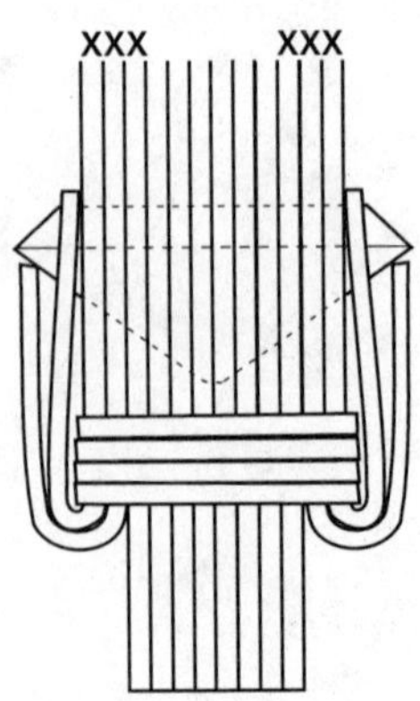

Diagram 102

G. After minor adjustments to the cradle wires, insert the stone as indicated in *Diagram 102.*

H. After the stone is inserted, make sure the point of the pavilion is well above the wrap wires, then spread three outside wires on either side of the group of 12 wires. They are marked with an x in *Diagram 102.*

I. Please keep in mind these diagrams are enlarged for clarity. See *Diagram 103* for clarification of the wires that have been spread.

J. One at a time, bend over each wire (in alternating diagonal order) across the cradle wires, starting with the outside wires first. Cut each one just long enough to wrap around the two cradle wires and crimp it tight. After each set of 4 is done, make sure the ring is still a size 9. *Diagram 104*

K. Make sure the ring is a size 9, then use the wide blade knife to turn up the outside shank wire on both sides. This will reduce the size of the ring by 1 1/2 sizes (to a 7 1/2). *Diagram 105*

L. *Diagram 105* shows the third shank wire turned up as well as the remaining 6 (group) wires spread. Just remember there are now four groups of 3 wires (two on each side of the ring). Each one is to be bent across the "corner" of the stone, beginning with the outside wire in alternating diagonal order until all are tied down neatly and in a tight parallel pattern, bent around the **THIRD** turned up wire and crimped snugly around it. *Diagrams 105, 106, & 106A*

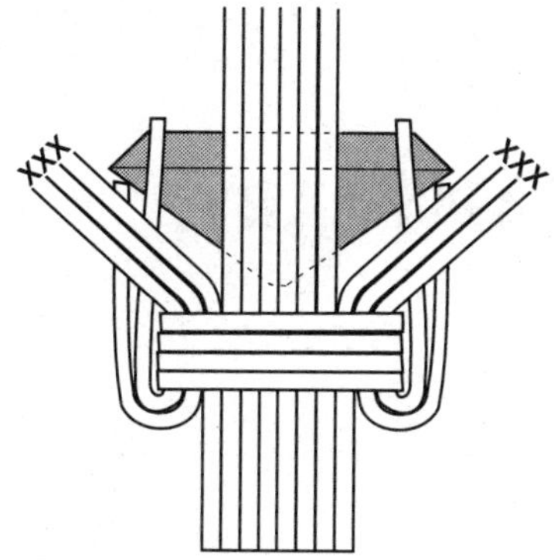

Diagram 103

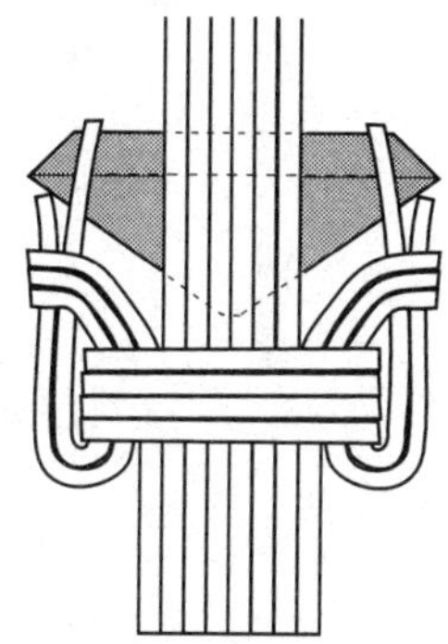

Diagram 104

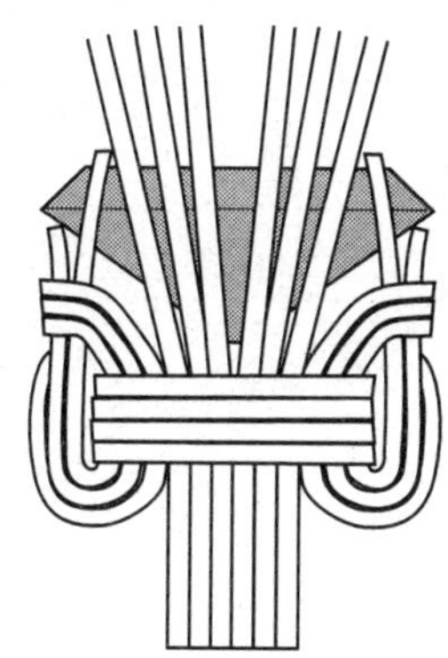

Diagram 105

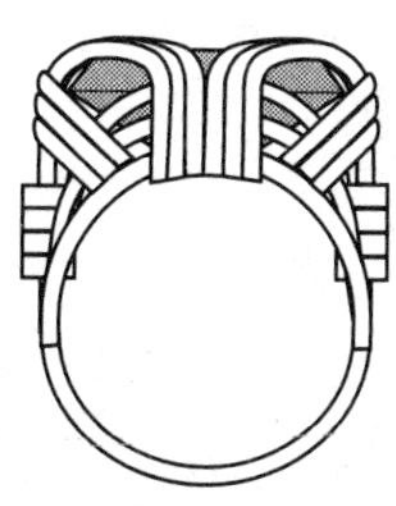

FRONT VIEW

Diagram 106

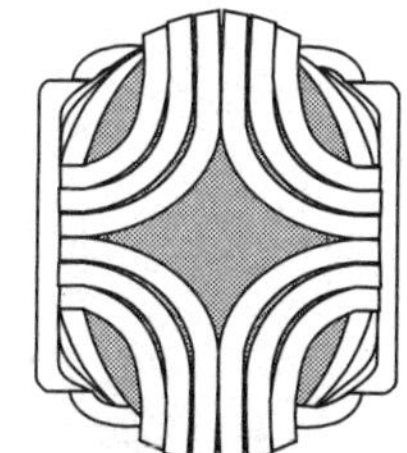

SHOWS ONLY
THE PLACEMENT OF THE
LAST 6 WIRES

TOP VIEW

Diagram 106A

SUMMARY:

This is a difficult ring to make, but with patience and perseverance and practice, it becomes easier. It is a very popular, attractive ring. Your skill level will determine what type of problems you will encounter, and how well you solve these problems will be determined by how much you have practiced.

Some of the problems you may encounter are:

Keeping the assembly tight
Keeping the wires in order
Keeping them flat and flush with each other
Bending properly and crimping neatly
Keeping it on size and
Keeping it symmetrical in the assembly process.

If you get too frustrated, leave it for a day or so (as you contemplate and ponder the importance of practice!) Remember, it can be done. There is a lot more to wirewrapping than just bending wire !!

#1 CAGE EARRINGS

To see these earrings at a glance, they really look complicated, but once the concept is understood, it is really a simple step by step procedure using a lot of common sense. For this system it is recommended to use thin wire. The models shown in this book have been done using mostly .028 square brass and copper wire. For a more delicate look, .025 can be used. It is also recommended to use square wire until you are very familiar with this system. The use of round wire also opens up a whole new dimension of variations (as you will soon discover).

A. Cut and clean six wires, (.028, square brass), 3" long. Two of them should be bent over on one end 1/2", 180 degrees. *Diagram 107* For large cages, six wires can be used for the cage and two for the wrap wires.

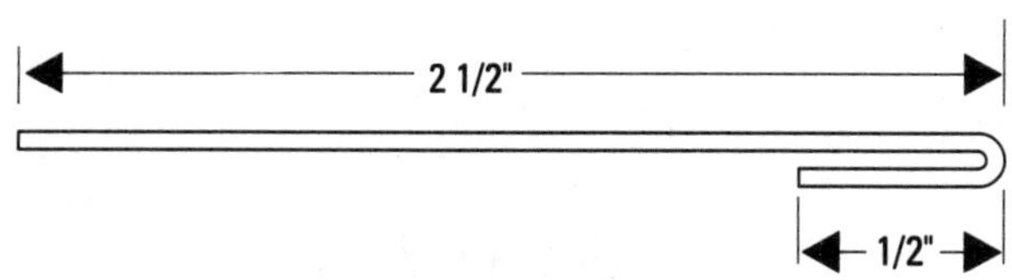

Diagram 107

B. Put the other four wires together (flat and parallel), and tie them together in the middle with one of the bent over wrap wires. *Diagram 108* Cut off excess wrap wire.

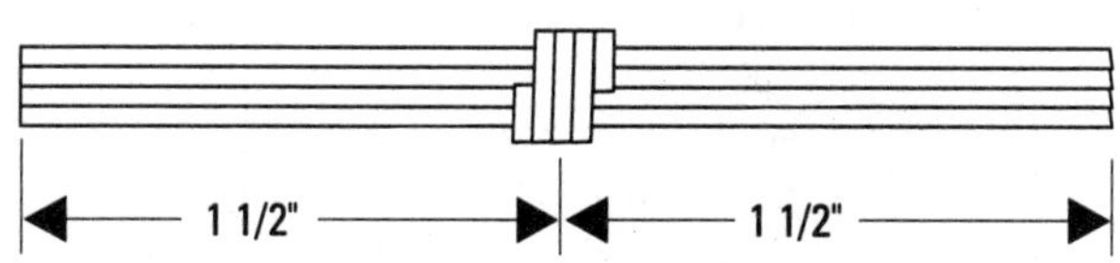

Diagram 108

C. Put the assembly on a flat surface making sure the ends of the wrap wires are facing away from you (not visible). **IT IS IMPORTANT TO DO THIS BECAUSE IT PUTS THE ENDS OF THE WRAP WIRE ON THE INSIDE OF THE CAGE.** Now measure out 5/ 8" from the **CENTER** of the wrap wire and mark with a felt tip pen. Do this to **BOTH** the left and right side of the wrap wire. *Diagram 109*

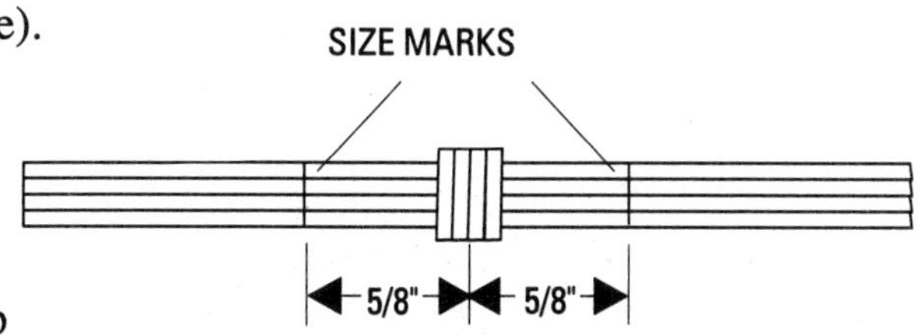

Diagram 109

D. Using the flat nose pliers, bend up the outer part about 30 degrees (bending at the 5/8" mark). *Diagram 110*

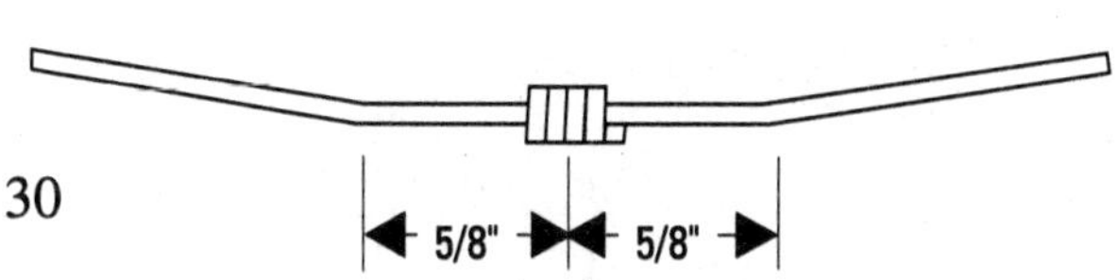

Diagram 110

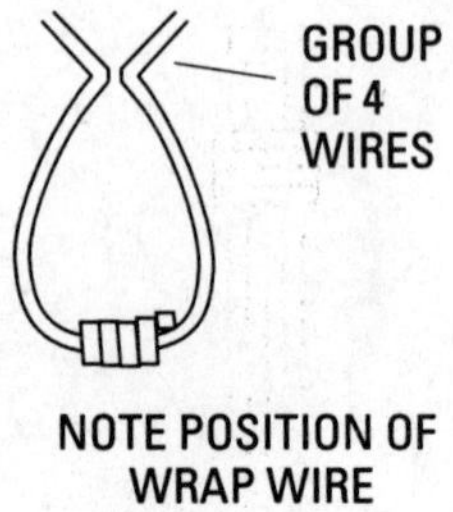

Diagram 111

E. Using a bolt (or a wooden dowel) 5/16" in diameter, bend the assembly around it, making sure that the ends of the wrap wire are against what you are wrapping it around (and not visible as you are bending). Bend it around until you get a teardrop shape. Remove the bolt (or dowel). See *Diagram 111* for accuracy.

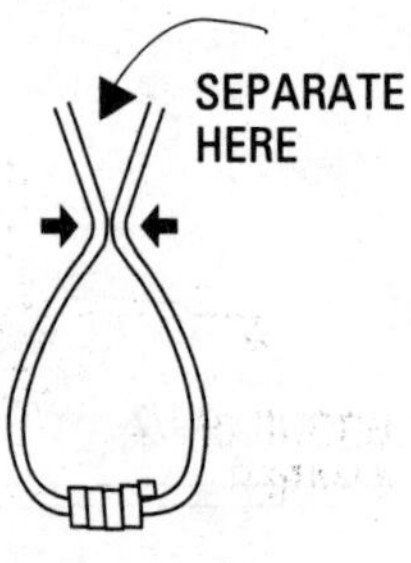

Diagram 112

F. Using the flat nose pliers, bend the group of 4 wires slightly toward each other. This is in preparation for tying them together with the other wrap wire. *Diagram 112*

G. Add the other wrap wire at the top of the tear drop. Wrap around 4 times and cut off excess. Separate each group of wires as indicated in *Diagram 113*.

Diagram 113

H. When these wires are separated, try to get the angle between them as close as possible to 90 degrees (a right angle) in preparation for the next step.
Make sure the second wrap wire wrap is **directly** above the bottom cage wrap.

I. Turn the assembly upside down at a 45 degree angle, place it and hold it on the vise as shown in *Diagram 114*. Hold at points A and B firmly, and tap **GENTLY** at point C to get a more rounded teardrop shape. While tapping **GENTLY,** take care not to get it off center (lopsided).

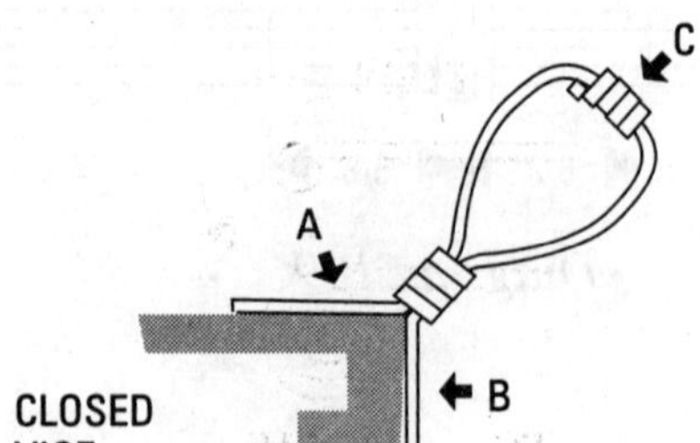

Diagram 114

J. The cage should now be a well-rounded small teardrop four wires wide. Do not separate these wires yet. *Diagram 115*

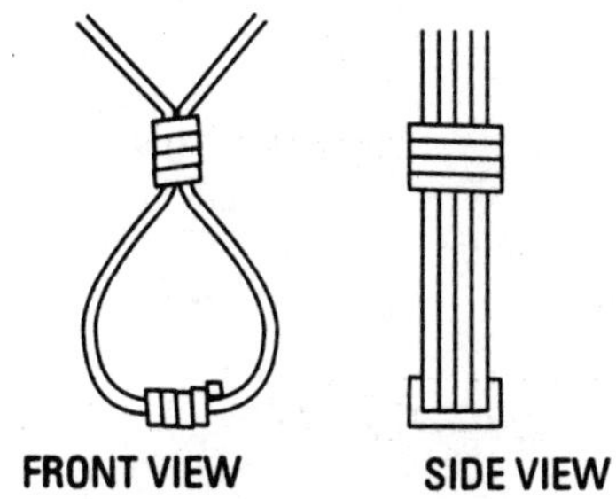

Diagram 115

K. With a wide blade pocket knife (or a kitchen knife), separate the two groups of 4 wires above the wrap wire. Leave one group straight up and down and the other group perpendicular to it. *Diagram 116*

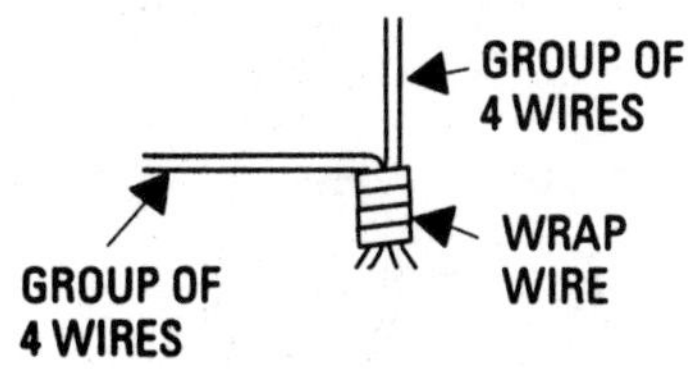

Diagram 116

L. The wires can be spread as indicated. **LEAVE TWO OF THE WIRES VERTICAL FOR A LOOP TO BE MADE** on which to put an ear wire. Trim all wires so they are 3/4" long. This diagram shows only the top view and is enlarged for clarity. *Diagram 117*

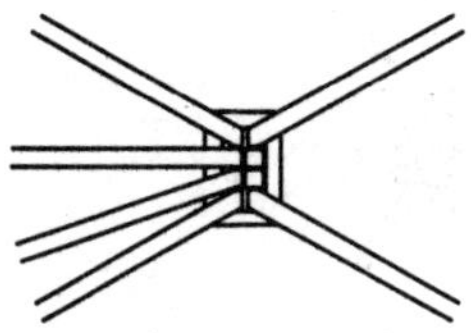

Diagram 117

M. These wires can be curled into flat discs. Begin these discs by using the small end of the round nose pliers and then continue on by gripping the disc in the edge of the pliers and pushing the wire into itself. *Diagram 118*

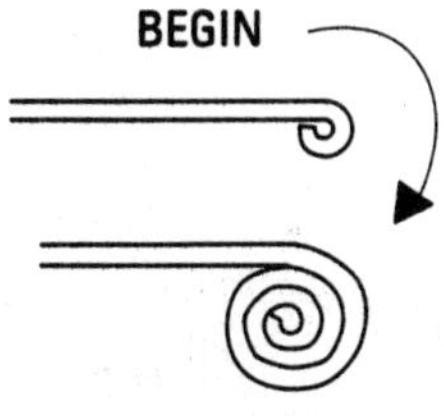

Diagram 118

N. These discs can be arranged in a neat pattern to cover the wrap wire on both sides of the cage.

Check color photo on cover or insert for finished piece.

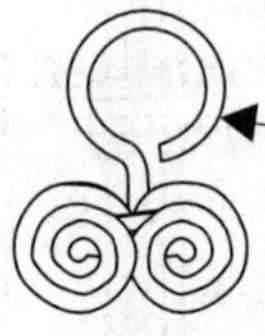

Diagram 119

O. The two remaining vertical wires can be bent as follows (using the larger portion of the round nose pliers). *Diagram 119*

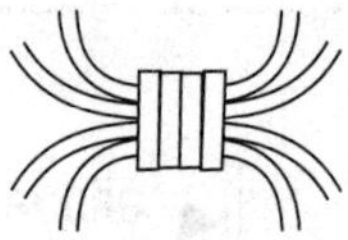

Diagram 120

P. The cage can now be spread as follows: Use the wide blade knife and make sure the blade is always parallel to the wire you are spreading. When spreading is completed, it should look like this (bottom view only). *Diagram 120* It shows the equal distribution of all 8 wires.

For further clarification, see color photo on cover or insert for finished piece.

SUMMARY:

There is a wide range of variations for this system, and after you become familiar with it, you are in for a real treat! Following next is a chart to be used as a guide to make a variety of different sizes, and it is in no way to be limited to what is stated here. This chart is merely suggestions for sizes. Yours may be a little different, depending on how neat your measure, bend or wrap. The columns in the chart are self-explanatory.

#2 SIZE CHART FOR CAGES

<table>
<tr><th>SIZE OF BEAD OR OBJECT TO BE CAGED</th><th>LENGTH OF WIRES ALL WRAP WIRES ARE 3"</th><th>NUMBER OF WIRES NEEDED</th><th>DISTANCE FROM CENTER OF BOTTOM WRAP TO SIZE MARK</th><th>BENDING TOOL TO BE USED</th><th>APPROXIMATE SIZE OF FINISHED CAGE</th></tr>
<tr><td>4mm bead</td><td>3 inches</td><td>4 plus 2 wrap wires</td><td>1/2 inch</td><td>1/4 " bolt</td><td>5/16 inch</td></tr>
<tr><td>6mm bead</td><td>3 inches</td><td>4 plus 2 wrap wires</td><td>5/8 inch</td><td>5/16" bolt</td><td>3/8 inch</td></tr>
<tr><td>9mm bead</td><td>4 inches</td><td>4 plus 2 wrap wires</td><td>3/4 inch</td><td>5/16" bolt</td><td>7/16 inch</td></tr>
<tr><td>12mm bead</td><td>5 inches</td><td>4 plus 2 wrap wires</td><td>7/8 inch</td><td>Size 1 on the ring mandrel</td><td>1/2 inch</td></tr>
<tr><td rowspan="4">Larger Beads or Small Mineral Eggs or Small Mineral Specimens</td><td>5 inches</td><td>4 (or 6) plus 2 wrap wires</td><td>1 inch</td><td>Size 3 on the ring mandrel</td><td>5/8 inch</td></tr>
<tr><td>5 inches</td><td>6 plus 2 wrap wires</td><td>1 1/8 inches</td><td>Size 5 on the ring mandrel</td><td>3/4 inch</td></tr>
<tr><td>5 inches</td><td>6 plus 2 wrap wires</td><td>1 1/4 inches</td><td>Size 9 on the ring mandrel</td><td>7/8 inch</td></tr>
<tr><td>6 inches</td><td>6 plus 2 wrap wires</td><td>1 1/2 to 2 inches</td><td>Size 13 1/2 on the ring mandrel</td><td>1 inch wide, 2 1/4 inch long</td></tr>
<tr><td colspan="6">FOR LARGE OR SMALL MINERAL SPECIMENS, I CHALLENGE YOU TO EXPERIMENT WITH YOUR OWN IMAGINATION. (IT'S A LOT OF FUN!)</td></tr>
</table>

VARIATION #1: CAGE WITHIN A CAGE, BOTTOM MOUNT

The variations look extremely unique when done in various combinations of wire (gold-filled and silver) (brass and copper) (gold and copper) (silver and brass) etc... These directions will show only how to incorporate these variations in the instructions already given. All other directions are the same.

A. Consult the cage chart for the measurements you need. **EXAMPLE:** These directions will be dealing with a caged 6mm bead inside of a larger cage, so you would cut four wires plus two wrap wires 3" long (.028, square brass or copper).

Tape four of the wires together at each of the **ENDS.**

Bend the other two wires over in preparation for using them as wrap wires. Do not wrap them yet. *Diagram 121*

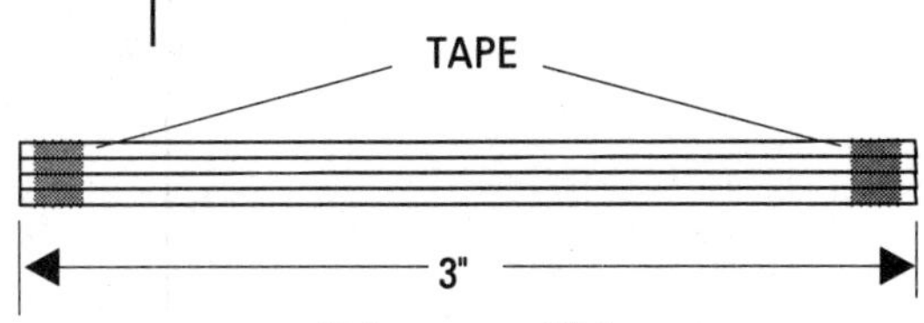

Diagram 121

B. Next, cut six wires 6" long, (.028 square) and **one** wrap wire 3" long. Tape together at the ends, like you did in *Diagram 121.* Do not use the wrap wire yet.

C. Mark the exact center of both "groups" of wire. Place the 6" group of wires on a flat surface, and then place the group of 3" wires on top of it so that the center mark is directly above the center mark of the group of 6" wires. Wrap the two groups together (one on top of each other, center marks together) with one of the 3" wrap wires. Make sure the cut ends (of the wrap wire) end up on top of the small group of 3" wires. *Diagram 122*

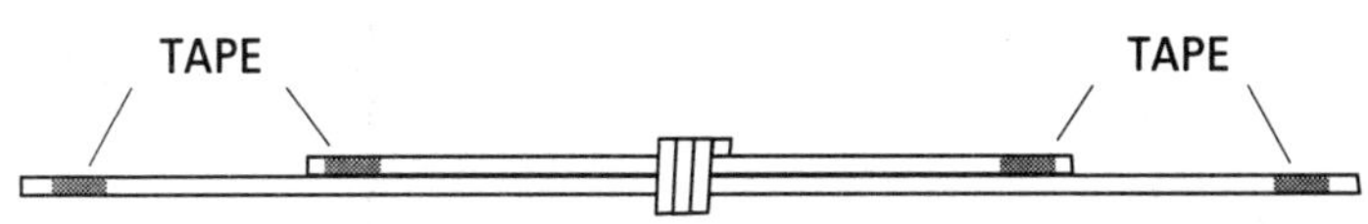

Diagram 122

D. Next, measure from the center of the wrap out on either side (on the 3" group of wires) 5/8" and mark with a felt tip pen.

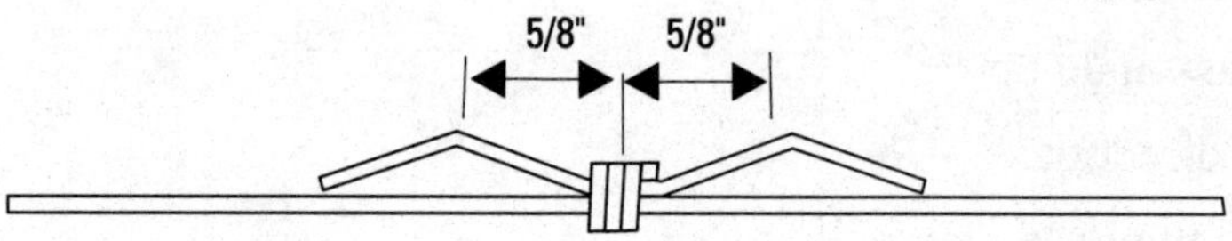

Diagram 123

E. Using a wide blade knife, pry up the 3" group of wires (very slightly) on either side of the wrap wires, then bend down slightly at each mark. *Diagram 123*

F. It is important to note at this point that you ignore the 6" wires until you are ready to make the large cage. You can make the smaller cage by working around them.

G. Insert a 5/16" bolt in the small cage assembly and bend **ONLY** the 3" wires around it.

FURTHER INSTRUCTIONS FOR COMPLETION OF THIS CAGE CAN BE FOUND BEGINNING WITH PARAGRAPH E, PAGE 72.

ONLY ONE EXCEPTION: PARAGRAPH L, PAGE 73. DO NOT LEAVE TWO WIRES VERTICAL FOR THE EAR WIRE SINCE THE CAGE IS SECURED AT THE BOTTOM. JUST COIL THE TWO WIRES WITH THE REST.

Remove the tape from the 3" wires.

H. Upon completion of this small cage, compare it with *Diagram 124*. The wires at the top of the cage can be cut to 1/2" and coiled (completely) into flat discs to cover the wrap wire.

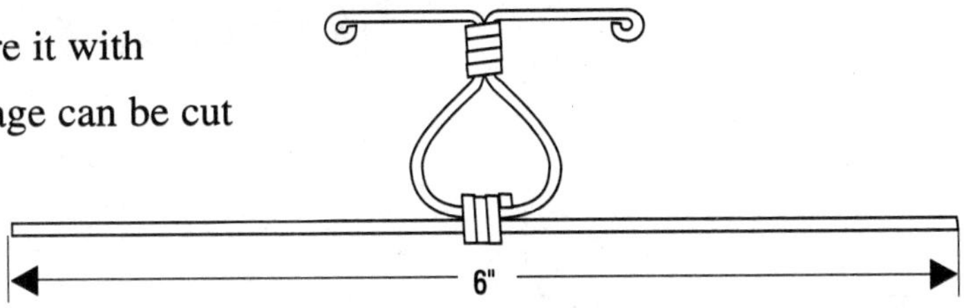

Diagram 124

I. Spread half of the wires and insert a 6mm bead, then spread the rest to complete the cage.

J. Now the 6" wires can be dealt with. From the center of the wrap wire, measure out 1 3/4" and mark, then using the flat nose pliers, bend down at the mark (on either side) all of the 6" wires that are still taped together.

K. From this point, you will need a delicate, gentle touch and a lot of concentration. The existing cage makes it impossible to bend and shape the outer cage around anything, so here is what's to be done:

L. Using the flat nose pliers and keeping in mind how to shape a bracelet (Volume 1, Chapter 4, paragraph U, page 53) do likewise with the 6" wires being careful not to disturb the attached cage. Give careful attention to keeping it centered around the small cage. It may take a little extra time to do this, but it can be done, and the end result is well worth it. Besides, this is what wirewrapping is all about! *Diagram 125*

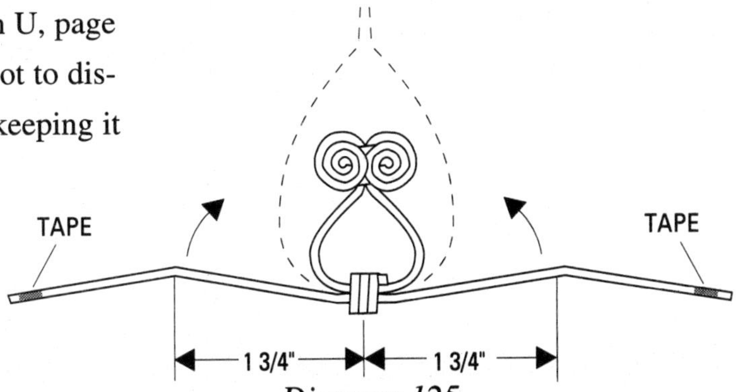

Diagram 125

M. Finish this cage by following paragraphs G through P in Chapter 16, dealing with cage earrings. (pages 72 to 74)

VARIATION #2: CAGE WITHIN A CAGE DANGLING FROM THE INSIDE TOP

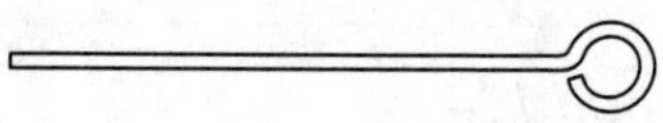

Diagram 126

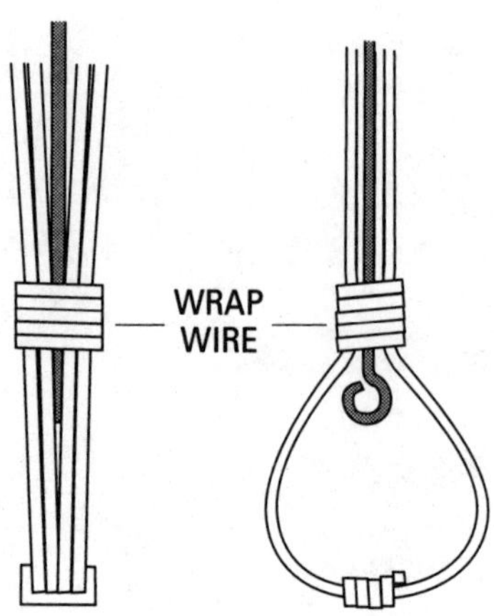

Diagram 127

A. Make a cage to accommodate a 6mm bead.

B. Make a larger cage as you did in Variation #1 except for the following addition: Before adding the top wrap wire, cut an additional 3" wire and on one end make a loop with the round nose pliers. *Diagram 126*

C. Insert this wire, with the loop pointing down (so it will be enclosed in the cage), as shown in *Diagram 127.*

D. Add the top wrap wire, and pull the inserted wire (with the loop) up tight against the roof of the cage. Wrap some of the surplus twice around and above the wrap wire. Cut off the excess. Continue making the larger cage by following paragraphs H through P, Chapter 16, pages 72 to 74.

E. Make sure the smaller cage is completed, and hang it from the loop in the larger cage. Make sure both loops are closed securely. The suspended cage should not touch the larger cage.

F. Spread the 6 wires (really 12) evenly with a wide blade knife. Variation #2 is now complete.

(See cover or insert for this completed item).

If you don't want the inner cage to dangle, then omit paragraph C. Do not make a hanging loop on the smaller cage. Leave it straight and incorporate it into the wires before you wrap the top of the large cage.

VARIATION #3: TANDEM CAGE EARRINGS

This type of earring is growing in popularity. Double or even triple tandem styles are extremely popular. From a feminine point of view, I guess it all depends on how tall you are. There are two ways to provide the connections for putting these earrings together. You can use the round nose pliers to add a jump ring to the bottom of the cage (it must go through the cage, parallel to the wrap wire already there), or if you are less experienced, you can add the jump ring while the cage is being made. Make it separately and slip it on the assembly just before the top wrap wire (of the cage) is added.

VARIATION #4: HINGED CAGES

These cages offer a great deal of versatility because of the advantage of being able to instantly change what is inside to match what you are wearing. The hinged cage can be difficult, and I would advise that you have a lot of experience as a wirewrapper before attempting to make one, or you will become needlessly frustrated. **THE KEY TO THE HINGED CAGE IS THE USE OF ROUND WIRE.** The directions that follow are for a **LARGE** hinged cage which offers a minimum of frustration. With practice, smaller ones can be made.

A. Cut six round wires 6" long (.028 or .032).
Cut three square wires 3" long (.028). Since these are wrap wires, bend over one end of them like always.

B. Using one of the 3" square wires, tie together the six round 6" wires in the center as indicated in *Diagram 128.*

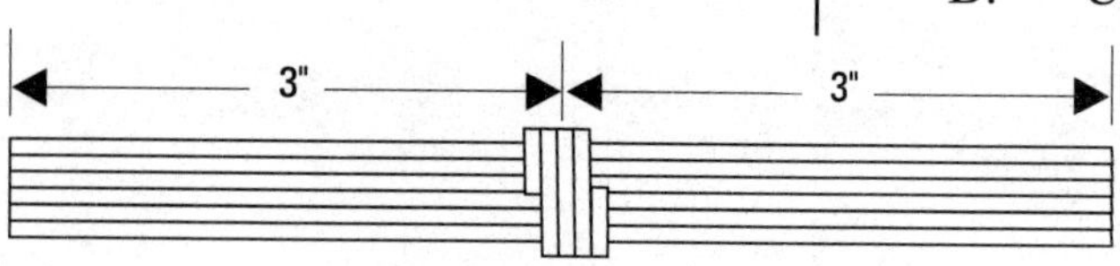

Diagram 128

C. During this entire procedure, be sure that the cut ends of **ALL** wrap wires are facing the same way.

D. Holding this group of wires firmly between the thumb and forefinger of the right hand, separate the wires on either side of the wrap wires into groups of 3 wires each and bend slowly and **GENTLY** by hand as in *Diagram 129.* **BOTH SIDES SHOULD BE AS IDENTICAL TO EACH OTHER AS POSSIBLE.**

Diagram 129

E. Use the flat nose pliers to bend each individual wire into the necessary straight stem on each group (marked with an X) of wires. *Diagram 129*

F. Each of these groups can now be wrapped with the other square wrap wires. Make sure they are extremely tight. *Diagram 130*

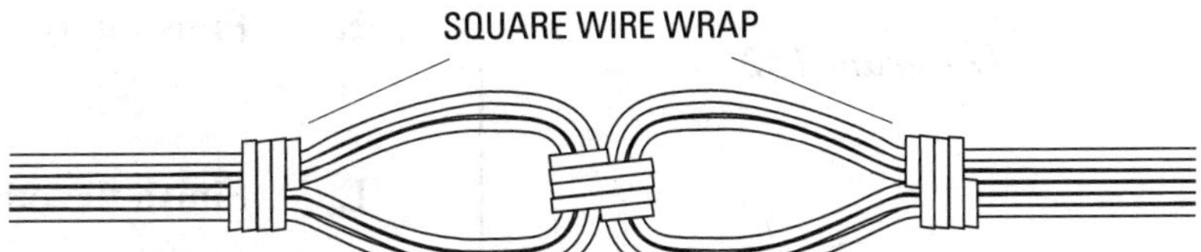

Diagram 130

G. Using the size 11 or 12 portion of the ring mandrel, bend each assembly separately around it as shown in *Diagram 131*. Before this is done, you may want to bend back the straight stem on each side just the way it is done when making the cages. Make sure the bend is at the lower end of the wrap wire.

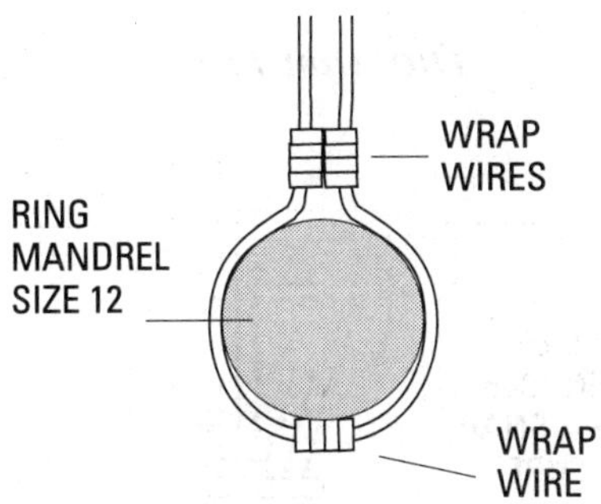

Diagram 131

H. Try to do this as neatly as possible--size is not that important--just keep it neat, tight and symmetrical. It should open and close easily at this point. Do not spread the wires to form a cage yet.

I. From this point, the diagrams will show only the top portion of this assembly. There are 6 wrapped wires perpendicular to the round part of the cage at the top. The diagrams will be dealing only with these wires--first one side and then the other. Remove the assembly from the mandrel and proceed.

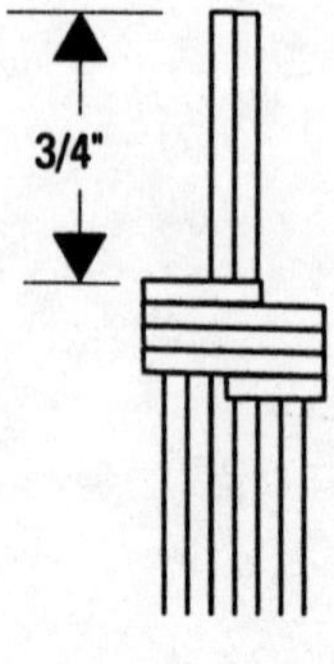

Diagram 132

Diagram 133

Diagram 134A

SIDE VIEW
Diagram 134B

J. Leave the two center wires as indicated in *Diagram 132*. Cut off the two outer wires on either side of them at the wrap wires.

K. Finish it as indicated in *Diagram 133.*

L. Finish the other side as indicated in *Diagrams 134A & 134B.*

M. It will be necessary to make minor adjustments. The large loop in *Diagram 133* should mesh between the large loop in *134B* and the snap closure loop also in *134B.*

N. Close the cage and adjust accordingly, using mostly the chain nose and round nose pliers.

O. While the round nose pliers are still in both large loops of the closed cage, spread the wires evenly so they form a sphere. The snap closure should fit over (just barely) the wrap wires on the other half of the cage to hold it closed. A neck chain through both loops should also prevent it from coming open while being worn.

DO NOT PULL ON THE CAGE WIRES TO OPEN THE CAGE. PRY GENTLY AT THE WRAP WIRES TO OPEN IT. Always snap it shut at the wrap wires **(GENTLY! !**) The cage is now complete except for whatever you want to put in it.

Diagrams 135A & 135B show both side views of the closure assembly.

LOOP SIDE
Diagram 135A

SNAP CLOSURE SIDE
Diagram 135B

SUMMARY:

To make this closing work properly, it takes a lot of patient minor adjustments on your part. Motivation depends on what kind of wirewrapper you want to be and how nice you want your creations to look.

After the cage is opened and closed a few times, you may have to adjust slightly the wrap wire at the bottom of the cage to make sure it is not tilted. Make sure it is level after each opening and closing.

A **GENTLE** push will suffice. Enjoy!!!

#1 ONE PIECE TRIPLE CAGE PENDANT

This is another variation of the cage. Once again your only limit is just how far your imagination will lead you. These instructions will be for a triple cage; however, more can be incorporated in a single unit. Any size can be made. This particular cage is 3 1/2" long when finished. For other lengths, use more or less wire.

A. Cut six wires 10" long. Either brass or copper .028 square. Cut three wires 3" long of the same kind and size and bend them over to be used as wrap wires.

B. Clean and straighten the wires, put them parallel to each other and wrap them together (in the center) with one of the 3" wrap wires. See *Diagram 136.*

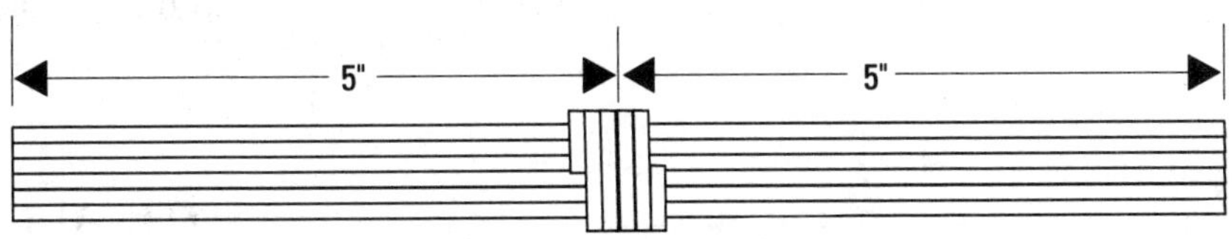

Tape each end as in *Diagram 138 A.*

Diagram 136

C. From the center of the wrap, measure out on each side 1 1/2" and mark. Make sure the cut ends of the wrap wires are facing down (away from you) before you mark. *Diagram 137*

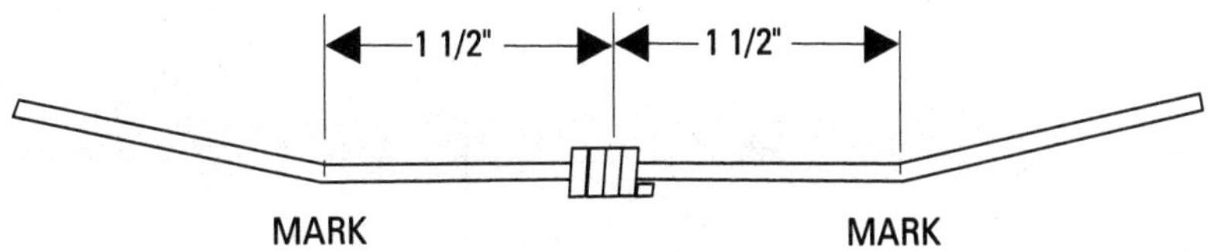

Diagram 137

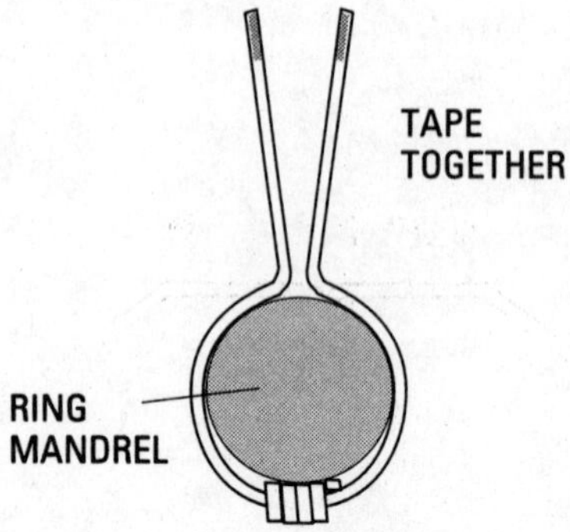

Diagram 138A

D. Place the assembly on a ring mandrel and bend around the size 12 1/2 to 13 making sure the cut ends of the wrap wires are on the inside. *Diagram 138A*

E. Remove the assembly from the mandrel, press together to form a circle, then secure with a wrap wire four or five times around. *Diagram 138B*

Diagram 138B

F. Repeat this process two more times, following the measurements given in *Diagrams 139 & 141*. The cages should be made progressively smaller.

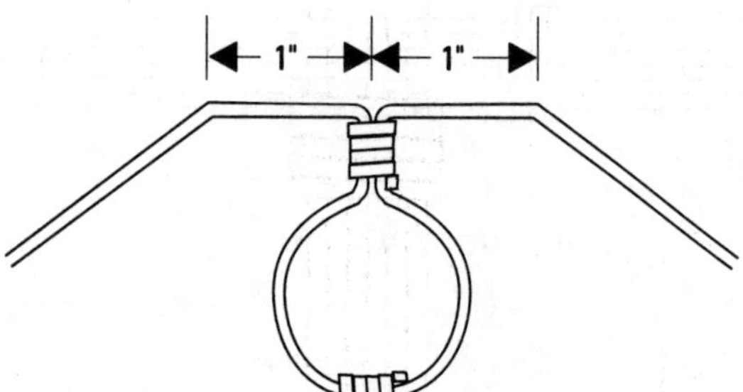

Diagram 139

NOTE: ALL CUT ENDS OF THE WRAP WIRES MUST BE ON THE SAME SIDE OF THE FINISHED PENDANT.

G. Finish the third cage by bending it around a 1/4" bolt and complete the top wrap wire as you did in *Diagram 140.*

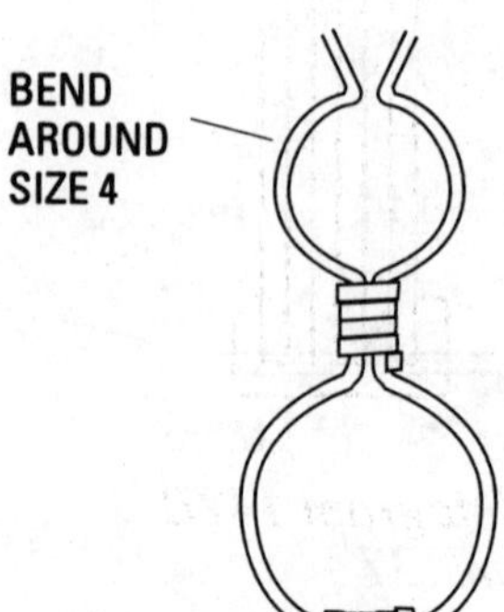

Diagram 140

H. *Diagrams 142A & 142B* are the triple cage pendant before the cage wires are spread. *See color photo on cover or insert for finished pendant.*

I. To finish this pendant at the top, see paragraphs L, M, N, O, and P in Chapter 16, dealing with cage earrings, pages 73 and 74.

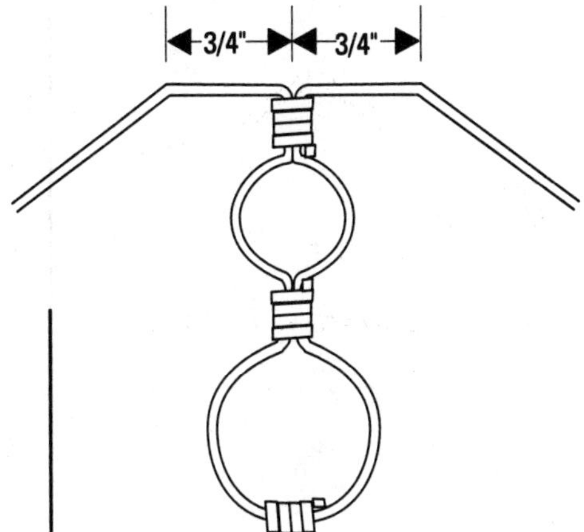

Diagram 141

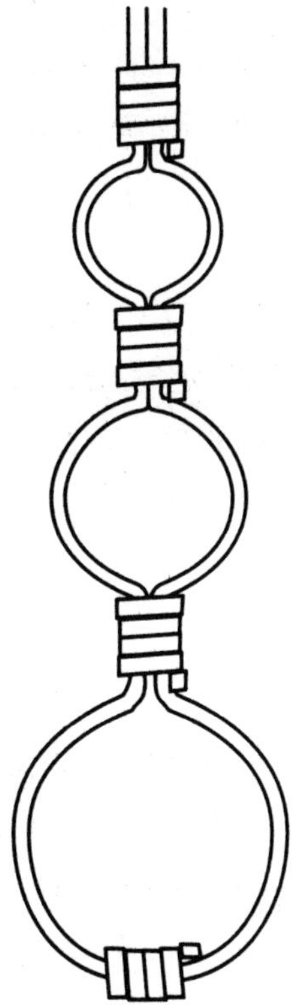

Diagram 142A

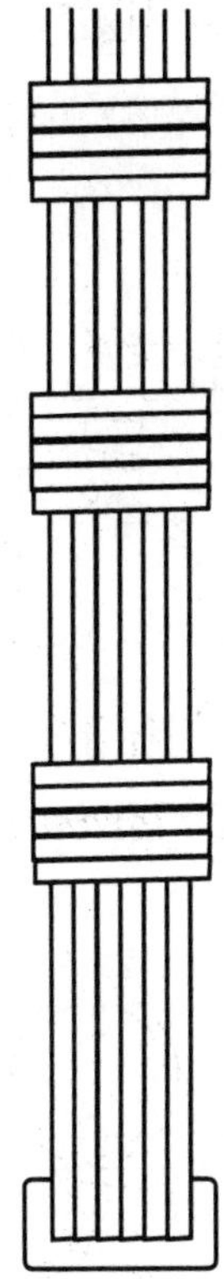

Diagram 142B

#1 EARRINGS

This system generates a lot of interest and excitement because of its versatility. It can be incorporated into earrings, pendants, a lapel pin, or with a little imagination, even into a bracelet. Spin off variations from this basic wrap really boggle the mind as one imagines an infinite number of wire-only jewelry sprinkled with a few beads. This chapter will only scratch the surface of what can be done when one really puts the imagination to work. These directions will show how to make a finished earring approximately 1 1/2" long (not including the loop for the ear wire). Square or round, brass or copper wire may be used. For that necessary delicate look, .025 or .028 wire should be used.

A. For making **two** earrings, you need to cut as follows:
Two wires, 6" long
Two wires 4" long
Two wires 2" long
Two wires 3" long (for wrap wires) These **MUST** be square.
You will need four beads, (6mm), two for each earring.
Keep in mind these wires are for two earrings. If these are made together, step by step, they will be closely identical to each other.

B. Tie together a 6", 4", and 2" wire with one of the 3" wrap wires as indicated in *Diagram 143*.
Make sure the wrap wire is very tight.

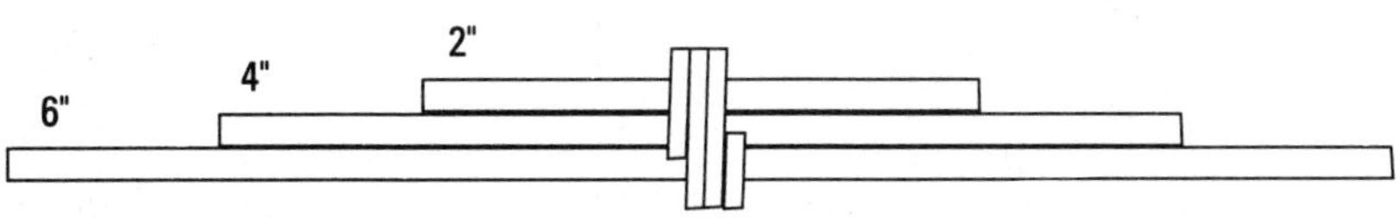

Diagram 143

C. Do likewise with the other set of wires.

D. Note the ends of the wrap wire. It is **IMPERATIVE** as you proceed with each step that **ALL CUT ENDS** are on the same side, and with the addition of the ear wires, you will have a left earring and a right earring. Further information will follow later.

E. Beginning with the 2" wire, bend each side up and try to keep it well-rounded. Add one bead to each wire (bent up). *Diagram 144*

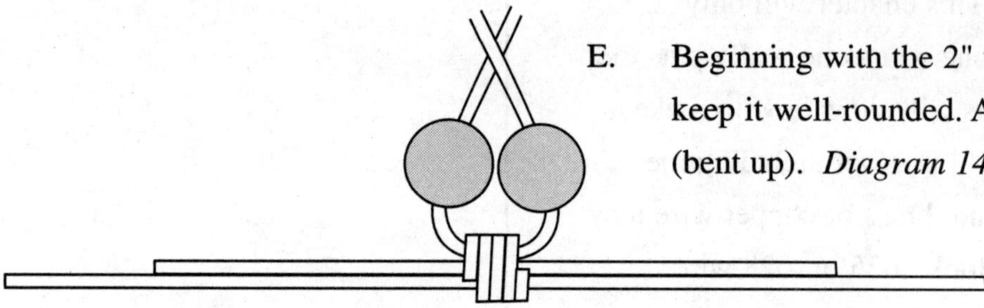

Diagram 144

F. For completion of this project, see *Diagram 145*. It's just a matter of shaping and bending the remaining wires and tying them down.

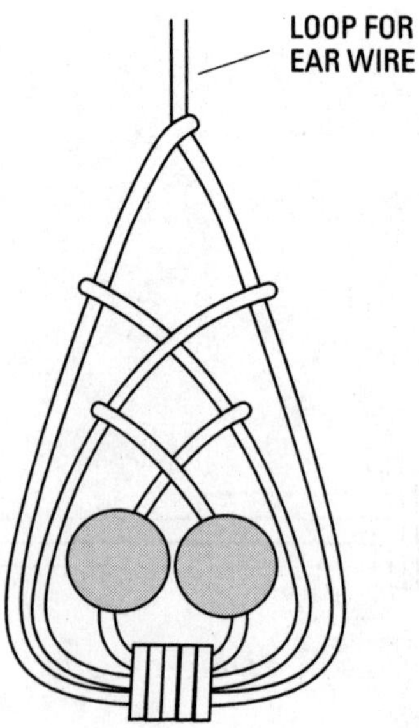

Diagram 145

G. **Note:** All cut ends are on the reverse side.
Center wire is the two inch wire.
Middle wire is the four inch wire.
Outside wire is the six inch wire.
Add ear wires in the opposite direction to accommodate left and right side.

See *Diagram 145A* for detail of tie-downs.

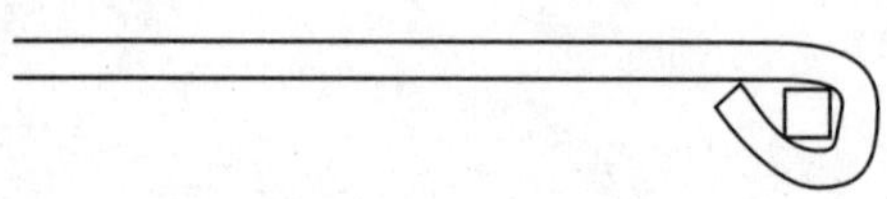

Diagram 145A

SUMMARY:

This earring can be made any desired size with any number (or size) bead. Using longer lengths of wire, more beads, or drilled stones can be added for a pendant. Instead of tying three wires together, put six wires together, and bend the other three in the opposite direction for a different design.

You can also experiment with this system in reference to a bracelet-**THINK BIG**- Your only limit is your own imagination.

CHAPTER 20

EPILOGUE

A rockhound is always looking for new, untouched areas to explore in his never ending quest for new and better specimens for his collection. He is always ready to go anywhere, anytime, to exert any effort, all for the glory of adding another one of nature's magnificent creations to his own private world of joy.

Such is the pursuit of a wirewrapper. There is no greater joy, other than mineral collecting, than to pursue a thought through the complex channels, crevasses, and ravines that wind endlessly within the matrix of one's mind. With tempered and patient thoughts, one can intellectually "chisel" away at an idea until the beautiful embryonic gem of a concept can be brought from vague abstraction to full reality in the form of a new ring, pendant or whatever you desire it to be.

Such is the joy of finding deep within yourself a treasure that you never suspected was there. It has been there for a long time waiting to be discovered. All of us have very special treasures within ourselves. With quiet, patient thoughts we can find these treasures. How many we find depends on how deeply we dig within ourselves. There is no limit to this potential. How far you want to dig depends on what you really want out of life and what you really want out of wirewrapping.

Many times I have been asked at shows, or during wire wrap classes, and even more so after the publication of Volume 1, *MOODS IN WIRE*, "Aren't you afraid of creating your own competition?" "Why are you giving away all your trade secrets?" These questions tend to answer themselves. As one becomes more involved in wirewrapping, the answer becomes more apparent. Sure, there is some duplicity in a few systems, but in the long run, the copycat artist quickly discovers he cannot duplicate someone else's personality. There is a very distinct difference among all wirewrappers. You may want to adhere to the motto:

DARE AND DISCOVER. CREATE, DON'T COPY!

(P.S. You more than double your rewards by sharing.)

Watch for the next surprising volume in this wirewrapping series coming out soon.

Less expensive materials and equipment can be easily obtained in many ways. Go to your local library and check out the Lapidary Journal Magazine for various catalogs in the jewelry trade.

Another good source for supplies is your local craft store, your local hardware store, and your local electrical supply store.

The best source for pliers is the ***Euro Tool Company*** *c/o your local dealer.*

Attend your local gem and mineral shows for further ideas.

NOTES

Notes

Notes

Notes

Notes